3301232805

D1806165

ROYAL COMMISSION ON ENVIRONMENTAL POLLUTION

CHAIRMAN: SIR JOHN LAWTON CBE, FRS

Twenty-seventh Report

Novel Materials in the Environment: The case of nanotechnology

Presented to Parliament by Command of Her Majesty
November 2008

Cm 7468   £ 26.60

PREVIOUS REPORTS BY THE ROYAL COMMISSION ON ENVIRONMENTAL POLLUTION

ISBN: 9780101746823

Royal Commission on Environmental Pollution

Twenty-seventh Report

To the Queen's Most Excellent Majesty

MAY IT PLEASE YOUR MAJESTY

We, the undersigned Commissioners, having been appointed 'to advise on matters, both national and international, concerning the pollution of the environment; on the adequacy of research in this field; and the future possibilities of danger to the environment';

And to enquire into any such matters referred to us by one of Your Majesty's Secretaries of State or by one of Your Majesty's Ministers, or any other such matters on which we ourselves shall deem it expedient to advise:

HUMBLY SUBMIT TO YOUR MAJESTY THE FOLLOWING REPORT.

"… for I was never so small as this before, never!"

Lewis Carroll, *Alice in Wonderland*, 1907

"Technology … is a queer thing. It brings you great gifts with one hand, and it stabs you in the back with the other."

C.P. Snow, *The New York Times*, 1971

More information about the current work of the Royal Commission can be obtained from its website at http://www.rcep.org.uk or from the Secretariat at Room 108, 55 Whitehall, London SW1A 2EY.

iv

Contents

Chapter 4

THE CHALLENGES OF DESIGNING AN EFFECTIVE GOVERNANCE FRAMEWORK

Chapter 5

SUMMARY OF RECOMMENDATIONS

REFERENCES

APPENDICES

FIGURES

INFORMATION BOXES

TABLES

Chapter 1

INTRODUCTION AND OVERVIEW

NOVEL MATERIALS

1.1　The discovery, development and deployment of novel materials have always been significant factors in the development of human civilisation. Prehistoric and historical epochs are even named according to the new materials (or new uses of materials) that were successively introduced and entered into common use during what we know as the Stone Age, Bronze Age and Iron Age.

1.2　In later eras, new materials have been closely associated with radical change. The development of paper was as important as the printing press in revolutionising communications. The introduction of gunpowder into Europe transformed warfare. In more modern times, gas lighting only became demonstrably superior to oil and candles with the introduction of the gas mantle, composed of novel materials such as thorium and cerium oxides. A hundred years ago electric filament lamps were made possible by other novel and fairly unusual materials, osmium and tungsten. More recently, fluorescent strip lights and compact high efficiency lights use once-novel phosphors to convert the UV produced by the electrical discharge into visible light.

1.3　Regardless of their novelty, materials are fundamental to all areas of technology and economic activity. Manufacturing and construction are entirely dependent on materials, and materials technology affects most economic activities.

1.4　The Royal Commission's decision to study novel materials was initially motivated by two kinds of concern. First was the potential for releases to the environment arising from increasing industrial applications of metals and minerals that have not previously been widely used. Second was the embodiment of nanoparticles and nanotubes in a wide range of consumer products and specialist applications in fields such as medicine and environmental remediation. As our inquiry progressed, it soon became clear that the bulk of evidence that we were receiving focused on the second of these issues.

APPLICATIONS OF NOVEL MATERIALS

1.5　Novel materials and new applications for existing materials are continually being developed in university and commercial laboratories around the world. They are intended either to improve the performance of existing technologies, such as fuel additives to improve the energy performance of cars, trucks and buses, or to make new technologies possible, such as MP3 players and mobile telephones which use trace quantities of exotic minerals. Novel materials are used under controlled conditions in industrial processes to make everyday objects. They are also incorporated in products which find their way into daily use.

1.6　Novel materials include a wide range of industrial products such as polymers, ceramics, glasses, liquid crystals, composite materials, nanoparticles, nanotubes and colloidal materials. In turn, these kinds of materials may be used in a wide range of applications including energy generation and storage, engineering and construction, electronics and display technologies, food packaging, and environmental and biomedical applications.

1.7   In the field of energy technology for example, the development of more efficient engines, advanced solar photovoltaics, improved batteries and hydrogen storage all offer opportunities for the potentially widespread application of novel materials. Diesel engines are said to be made more efficient by the use of fuel additives, such as cerium oxide. Jet engines can burn fuel at much higher temperatures when rhenium is added to alloys used in their construction. Conductive organic polymers, inorganic semiconductors such as cadmium selenide (in both bulk and nanoparticulate forms) and fullerenes are of interest to manufacturers of solar cells. Various novel lithium compounds are being investigated to achieve improvements in the cathodes of lithium ion batteries found in numerous portable electronic devices, including laptop computers and mobile phones. Hydrogen could be used as an alternative to electricity as an energy source and storage medium. But hydrogen storage as gas or liquid currently presents problems that could potentially be overcome by using inorganic metal hydrides of light elements (along with platinum, palladium, nickel or magnesium as catalysts) or by absorption in high porosity materials with large surface areas, such as nanotubes. There is a similarly wide range of potential applications in many other fields.

1.8   Novel materials are developed in response to a number of different drivers, including the requirement for a specific or improved functionality, increased efficiency, and the need to find substitutes for raw materials that are in short supply or have been found to have adverse effects on the environment or human health. An example of where safer substitutes for existing materials are desirable is the replacement of lead solder in electronic devices. In some cases, the discovery of novel functionality (the ability of a material to behave in a certain way or to 'do' something) actually drives a search for profitable applications.

1.9   The improved efficiency and functionality of novel materials can bring tangible environmental benefits, such as those offered by the development of photovoltaics, fuel cells and lightweight composites for cars and aircraft. In all cases, it is unlikely that new materials will be adopted, even in critical areas such as low-carbon energy technology, if the price is too high.

1.10   An example of materials innovation to reduce costs is the search for alternatives to the use of silicon transistors in liquid crystal displays (LCDs). While this technology is well understood, it remains costly and energy intensive, and manufacture of the materials involves the use of highly corrosive chemicals. Conducting polymers, transparent conducting oxides, silicon nanorods and carbon nanotubes are all being explored in the development of printing technologies that could achieve large display area capabilities, high processing speeds and low energy input.

1.11   Price may be only one of a number of constraints on the development and deployment of novel materials. For example, the scarce supply of some elements, such as indium, means that there may not be sufficient availability to realise the potential benefits on a substantial scale.

1.12   When scarce new materials are used in very small quantities, for example as dopants in electronic equipment, the feasibility and cost effectiveness of recycling them is diminished so that increasingly they will be released into the environment.

1.13   Some novel materials of concern are themselves already the subjects of searches for substitutes on either cost or health grounds. Cadmium, selenium and indium used in photovoltaics, and tellurium, bismuth and lanthanum in magnetic storage devices are all considered toxic. While they appear to pose no threat in use, they require careful handling in manufacture (especially to avoid contamination of wastewater streams) and during end-of-life recycling or disposal.

DEFINITIONS OF NOVEL MATERIALS

1.14   The first question that we faced was how widely we should cast the net of 'novel materials'. Clearly we did not wish simply to reproduce our Twenty-fourth Report, *Chemicals in Products*.[1] In embarking on this report, we initially found it useful to distinguish four types of novel materials:

- new materials hitherto unused or rarely used on an industrial scale, such as certain metallic elements (e.g. rhodium, yttrium, etc.) and compounds derived from them;

- new forms of existing materials with characteristics that differ significantly from familiar or naturally-occurring forms (e.g. nanoforms of silver and gold that exhibit significant chemical reactivity, enhanced biocidal properties or other properties not manifest in the bulk form);

- new applications for existing materials or existing technological products formulated in a new way, which may lead to substantially different exposures and hazards from those encountered in past uses (e.g. the use of cerium oxide as a fuel additive); and

- new pathways and destinations for familiar materials that may enter the environment in forms different from their manufacture and envisaged use (e.g. microscopic plastic particles arising from mechanical action in marine ecosystems).[i]

1.15   Despite the breadth of these definitions, most of the evidence that we received focused on nanomaterials – particles, fibres and tubes on the scale of a few billionths of a metre (Chapter 2). The emphasis on nanomaterials may have been due to a tendency among those offering us evidence to equate 'novelty' with 'revolutionary' change. It might be the case where research builds incrementally on existing knowledge and the new properties are not altogether unexpected, that their creators do not consider the results to be 'novel materials'. However, where there are revolutionary changes in the properties and levels of understanding of a material then it may be more likely to be considered 'novel'. Hence, it is perhaps unsurprising that many of the materials about which we received evidence were nanomaterials, many with truly novel properties as described in Chapter 2.

1.16   The properties of a novel material can arise from two key factors: first, the chemical composition of the material and second, its physical size and shape. As scientists exert ever more sophisticated control over molecular level organisation, the morphology of materials is becoming increasingly important. The example of gold illustrates how physical properties can change the chemical properties of a material. In its natural bulk form, gold is famously inert. Naval uniform buttons

---

i   Another approach might be to consider materials referred to by laws and regulations as 'new' or 'novel', for example under the toxic substances legislation of the USA (Toxic Substances Control Act, TSCA) or the European chemicals regulation (REACH). Here 'novelty' may have little basis in science and is often defined by whether or not a substance is on an existing regulatory database (e.g. the European Inventory of Existing Chemical Substances, EINECS).

were often gilded, in part to resist corrosion from salt air (Army buttons being more often made of ungilded brass). However, at a particle size of 2-5 nm, gold becomes highly reactive. The chemical composition of these two materials is identical: it is the different physical size of bulk materials and nanoparticles that accounts for their very different chemical properties.

1.17 The example of gold points to a consideration that has consistently guided us in our inquiry. *It is not the particle size or mode of production of a material that should concern us, but its functionality.* Indeed, we encountered several experts who observed that the focus of attention is switching from the size of particles to what they actually do. These experts predict that the term 'nanotechnology' will disappear within a decade or so. This reinforced our view that the key factors that should drive our interest in the environmental and human health issues surrounding novel materials are, indeed, their functionality and behaviour.

1.18 It would even be consistent with this emphasis on functionality to define a novel material as one whose effects on human and ecosystem health are currently not understood.[2] Of course, there are many materials that have been around for a long time whose toxicology is not fully understood. However, there are also whole new categories of materials currently being produced (particularly nanoparticles) for which toxicological and ecotoxicological data are entirely lacking.

1.19 An approach to the classification of novel materials that takes account of their functionality is employed by the Woodrow Wilson Center. It distinguishes four types:[3]

- **evolutionary materials:** Materials whose existing properties are enhanced or made more accessible or useable. Examples would include sophisticated metal alloys and engineered nanoparticles of metals and metal oxides, where increasing surface area and decreasing particle size affect bulk properties like reactivity and light scattering;

- **revolutionary materials:** Materials that are not an extension or evolution of familiar or conventional materials, but are distinct materials in their own right. Examples would include carbon nanotubes, fullerenes, dendrimers and quantum dots (2.5-2.8);

- **combination materials:** Composite materials where a combination of two or more components leads to unexpected or unconventional properties. These would include the use of carbon and metal oxide nanoparticles and nanotubes in composites, leading to changes in strength, conductivity and other physical and chemical properties. They would also include more complex nanoparticles where multiple components have been engineered into the final product, including smart nanoparticles for treatment of cancer and other diseases, and core-shell nanoparticles where outer layers of a different material have been added to alter functionality; and

- **materials with the potential for unanticipated and unusual biological impact:** Some new materials might behave predictably in the applications they are designed for, but present unusual and unanticipated health and environmental hazards. Novelty in this case comes from the potential to cause harm in unconventional ways. Within the bounds of current knowledge, this category encompasses most manufactured nanomaterials that are based on, or have the ability to release, low-solubility nanoscale or nanostructured particles into the environment. Some such particles may be capable of interacting with biological systems in different ways to those of larger particles. They may be small enough to cross biological barriers that are typically impermeable to larger particles. Others may be transported and accumulate in the environment in ways that are different from conventional materials. This category might also include materials with surface structures at the nanoscale that can potentially interfere with biological processes.

1.20 No single exhaustive taxonomy of novel materials has yet been devised. We believe it is unlikely that one is possible or even necessarily desirable. Each approach emphasises different attributes of the materials in question and their applications. *However, the functionality of the material, i.e. what it is designed to do and how it is capable of achieving it, appears to be the most robust focus for evaluating its potential environmental and human health implications.*

FUNCTIONALITY: SHOULD WE BE CONCERNED?

1.21 The environmental and public health implications of novel materials have attracted little attention from the public or policy-makers, with the exception of nanomaterials, which have been addressed in a number of reports on the broader topic of nanotechnology; a topic which, for a while, vividly captured the attention of the mass media on both sides of the Atlantic.

1.22 While there have been no significant events that would lead us to suppose that the contemporary introduction of novel materials is a source of environmental hazard, we are acutely aware of past instances where new chemicals and products, originally thought to be entirely benign, turned out to have very high environmental and public health costs. The list includes: asbestos, a life-saving fire retardant and valuable insulator that causes serious lung disease; chlorofluorocarbons, which were thought to be entirely harmless in a variety of applications including refrigeration, insulation and electronics, but turned out to have enormously damaging consequences for the atmosphere; tetra-ethyl lead, an anti-knocking compound in petrol which was injurious to the mental development of children exposed to exhaust fumes; or tributyltin, an antifouling paint additive used on ships' hulls which bore serious consequences for a range of marine organisms.[4] In light of such past experiences and recent research findings,[5] we note that the Environment Agency has recently taken the precautionary approach of classifying waste containing unbound carbon nanotubes as hazardous.[6]

1.23 There is a long history of adverse human health effects caused by occupational exposure to chemicals and inhaled dusts. Usually exposures need to be substantial and prolonged, as was the case for pneumoconiosis, the severe fibrotic lung disease associated with coal mining. However, high levels of exposure are not needed in the case of the highly malignant cancer mesothelioma associated with asbestos exposure, where the mineral characteristic of the fibre (diameter, length and persistence), as well as level and type of exposure, is a critical factor. Fortunately, with the exception of mesothelioma which has a lag time of many years, these diseases are progressively declining with the introduction of improved occupational hygiene and, in some cases, complete removal of the offending agent from use. In these cases, an appreciation of the cause and effect relationship is important so that appropriate safety measures can be implemented on the basis of validated toxicological testing.

1.24 However, such safety measures can only be introduced if the association between the substance in question and adverse health effects is known. A recent example that extends beyond the workplace is the discovery of the adverse pulmonary and cardiovascular effects of ambient air pollution particles from vehicle emissions. This emerged from careful population-based epidemiology, which is able to take account of confounding factors such as geographical location and socio-economic status. Although the underlying mechanisms are still not fully understood, the ability to derive exposure–response relationships between particles of a particular size and mass and human health effects has enabled robust air quality standards to be set to protect the public. Learning from this experience, if new materials are introduced it is essential that every effort is made to understand their toxicity profile in relation to human health and the wider environment.

1.25   It is a matter of concern that we were repeatedly told by competent organisations and individuals that we do not currently have sufficient information to form a definitive judgement about the safety of many types of novel materials, particularly many types of nanoparticles. In some cases, the methods and data needed to understand the toxicology and exposure routes of novel materials are insufficiently standardised or even absent altogether. There appears to be no clear consensus among scientists about how to address this deficit.

1.26   Experts seem to agree that there is considerable uncertainty about what kinds of environmental and toxicological effects might be expected. Will novel substances simply give rise to known effects but to a different extent when compared to established materials or might they give rise to completely new, as yet unknown, environmental effects? Current testing protocols are fairly coarse screening mechanisms which tend to pick up acute effects. Almost by definition, with novel materials there are virtually no data on chronic, long-term effects on people, other organisms or the wider environment.

1.27   Under current procedures it can take up to 15 years for a new testing protocol to achieve regulatory acceptance. Given the rapid pace of market penetration of novel materials and the products that contain them, existing regulatory approaches cannot be relied upon to detect and manage problems before a material has become ubiquitous.

1.28   Difficulties also arise because the form in which materials make their way into the environment might not be the same as that encountered during manufacture. Many free nanoparticles agglomerate and aggregate in the natural environment, forming larger structures that may have different toxicological properties to those exhibited by the original nanoform.

1.29   Most novel materials are used in factories or incorporated in products, but our inquiries suggested that very little thought has been given to their environmental impact as they become detached from products in use or at the point of final disposal. For example, little attention is paid to the ultimate fate of novel pharmaceuticals in the environment following elimination from patients.

1.30   Determining the fate of novel materials is vital when assessing the toxicological threat they pose. Nanomaterials are illustrative of the challenge. Techniques for their routine measurement in environmental samples are not widely available, nor are we currently able to determine their persistence in the environment or their transformation into other forms. Laboratory assessments of toxicity suggest that some nanomaterials could give rise to biological damage. But to date, adverse effects on populations or communities of organisms *in situ* have not been investigated and potential effects on ecosystem structure and processes have not been addressed. Our ignorance of these matters brings into question the level of confidence that we can place in current regulatory arrangements.

TRANS-SCIENCE, WORLD VIEWS AND THE CONTROL DILEMMA

1.31   The policy challenge posed by novel materials is a specific instance of the more general dilemma of how to govern the emergence of new technologies which, by definition, cannot be fully characterised with respect to their potential benefits and drawbacks. As such it is a classic case of what the American physicist Alvin Weinberg described as a 'trans-scientific' problem.[7]

1.32   Trans-scientific questions are those that can be posed in the language of science as questions of fact, but are in practice unanswerable by it. A classic instance is the question "Is it safe?", to which the answer must always be a matter of judgement and not of fact. Judgement is more difficult in situations where there is little or no consensus about what constitutes the evidence on which it might rest.

1.33   World views incorporate ethical values as well as ontologies (ideas about the nature of things). Scientists and regulators, as well as the wider public, invariably use world views to interpret data or other kinds of evidence. But where information is missing or evidence is ambiguous, people draw even more heavily on more general world views to inform their decision making. For example, those who believe that nature is maintained in a delicate balance are more likely to regard any discharge into the environment as a dangerous insult than those who see nature as robust and forgiving.

1.34   These contrasting world views are highlighted by various reports on nanotechnology published on either side of the Atlantic in the first half of this decade.[8] US reports tend to concentrate on the upside of nanotechnology, describing its potential in glowing, often Utopian terms. European reports tend to dwell more on potential dangers to health, environment and the social fabric. Yet there is no substantial difference in the scientific or technological data available to the authors of these reports. In new situations, individuals and institutions rely on their existing ideas and beliefs about risks and how they should be managed.

1.35   In gathering our evidence for this report it was clear to us that different organisations and individuals interpreted the same information, or lack of it, in very different ways, reflecting their broader interests and outlooks. We heard at least three distinctive approaches to the problem of the governance of novel technologies under conditions of what we consider to range from high uncertainty to profound ignorance.

1.36   One optimistic view was that no regulatory attention to novel materials could be justified unless and until there were clear indications that harm is being caused. Those expressing such a position were generally more concerned to forestall any unjustified regulatory intervention that might stifle innovation. A less optimistic version was the argument that any attempts to devise governance arrangements for novel materials should be 'risk based'. This usually means that the technology should be controlled only to the extent that there are clearly articulated (preferably quantified) scientific reasons for concern, and only then where the cost of risk reduction is deemed proportionate to the probability and extent of danger. Reasons for concern might include detection of empirical disease clusters, the articulation of theoretically plausible exposure pathways, or plant or animal disease mechanisms that might be associated with particular novel materials. At the other extreme was the view that novel materials should not be permitted until they had been given a clean bill of health, i.e. they had been demonstrated beyond any reasonable doubt to be safe.

1.37   We were not persuaded by any of these positions. The first assumes that nature is always benign until proven otherwise. As we have noted, history is replete with instances where such assumptions were shown to be flawed too late to avoid serious consequences. The second approach assumes that the state of the science is up to the job of detecting problems unambiguously and at an early enough stage to prevent widespread damage, which we have not found to be the case here. The third view would deny citizens and consumers the real lifestyle and health benefits that technologies based on novel materials might provide. In any case, we know that science can never definitively prove that something is safe.

1.38   Contemporary society is characterised by the accelerating pace of the proliferation of new technologies. Increasingly, it will be impossible to settle questions about the environmental and human health impacts of new materials consistently and in a timely fashion using traditional risk-based regulatory frameworks. The problem is exacerbated by the fact that in a technologically interdependent world, individual states cannot realistically exert the power to monitor and enforce rules governing the incorporation of materials in a wide range of products or their disposal.

1.39   We are faced with an instance of what David Collingridge described as the 'technology control dilemma'. As long ago as 1980,[9] he suggested that in the early stages of a technology we don't know enough to establish the most appropriate controls for managing it. But by the time problems emerge, the technology is too entrenched to be changed without major disruptions.

1.40   The solution to this dilemma is not simply to impose a moratorium that stops development, but to be vigilant with regard to inflexible technologies that are harder to abandon or modify than more flexible ones. Thus, key questions are how reversible is society's commitment to the technology and how difficult would it be to remediate if problems arose. Among the technical and social indicators of inflexibility are: long lead times from idea to application; capital intensity (such as investment in large plant and costly equipment); large scale of production units; major infrastructure requirements; closure or resistance to criticism; exaggerated claims about performance and benefits; and hubris. To this list we might add irreversibility, in the form of widespread and uncontrolled release of substances into the environment. According to this approach, the more of these indicators that are present, the more cautious we should be in committing ourselves to adoption of the technology.

1.41   These considerations of trans-science, world views and the control dilemma suggest that novel materials, like other emerging areas of technology, require an adaptive governance regime capable of monitoring technologies and materials as they are developed and incorporated into processes and products. An effective, adaptive governance regime will have to be capable of applying the indicators of technological inflexibility identified in the technology control dilemma to decide when to intervene *selectively* in areas where it deems that a material represents a danger to the environment or human health. While any kind of blanket moratorium does not seem appropriate, there may well be specific cases where it is necessary to slow or even hold up the development while concerns are investigated.

1.42 Such a governance regime would be consistent with and build upon a recommendation from the 2004 Royal Society and Royal Academy of Engineering report on nanoscience and nanotechnology[10] in relation to the governance of nanotechnology, which proposed the establishment of a "group that brings together representatives of a wide range of stakeholders to look at new and emerging technologies and identify at the earliest possible stage areas where potential health, safety, environmental, social, ethical and regulatory issues may arise and advise on how these might be addressed".

THIS REPORT

1.43 In preparing this report, our aim is to provide a framework for thinking about and addressing concerns about the impacts of novel materials. Hence, in Chapters 2 and 3, we explore the extent to which novel substances are currently being deployed, the plausible pathways by which they might enter the environment, their likely environmental destinations in use or disposal and the possible consequences of their release to those destinations. In Chapter 4 we go on to consider what arrangements would be most appropriate for the governance of emerging technologies under two conditions that pose serious constraints on any regulator. First is the condition of ignorance about the possible environmental impacts in the absence of any kind of track record for the technology. Second is the condition of ubiquity – the fact that new technologies no longer develop in a context of local experimentation but emerge as globally pervasive systems – which challenges both trial-and-error learning and attempts at national regulation.

1.44 Both new governance approaches and modifications to existing ones are likely to be called for. They will need to be rooted in ideas of adaptive management that require multiple perspectives on the issues. In the meantime, we emphasise that it makes little sense to frame the governance challenges in terms of whether industry, government, or citizens should be 'for' or 'against' nanomaterials or any other kinds of novel materials. It is the functionality of the material, not particle size or mode of production, which is critical for evaluating its potential impact on the environment or human health.

Chapter 2

PURPOSE, PRODUCTION AND PROPERTIES OF NOVEL MATERIALS: THE CASE OF NANOMATERIALS

INTRODUCTION

2.1   We concluded in Chapter 1 that novelty depends on the exploitation of properties of a substance, particularly to deliver new functionalities, not on the particular processes used to manufacture them or simply on their composition or size. In this chapter, we discuss more fully how functionality can determine both the uses to which a material can be put and its potential to cause harm to the environment and human health. We also look at the innovation system for nanomaterials, identifying the different actors and their linkages, and examining how this system can be used to help develop policy for the management of nanomaterials.

2.2   The behaviour of novel manufactured materials, particularly manufactured nanoparticles, should be seen in the context of the existence of naturally-occurring nanoparticles (2.7) to which the environment and organisms have been exposed for millions of years. Indeed, there have been long-standing uses of what we now recognise as nanomaterials, as illustrated by the Lycurgus cup, shown on the cover of this report. The Lycurgus cup is thought to have been made in Rome in the 4th century AD. The cup is the only complete example of a very special type of glass, known as dichroic, which changes colour when held up to the light. The opaque green cup turns to a glowing translucent red when light is shone through it. The glass contains tiny amounts of colloidal gold and silver, which give it these unusual optical properties.[1]

2.3   We pressed many witnesses and organisations on whether they had concerns about potential environmental and human health impacts of non-nanoscale novel materials which could not already be addressed through the current regulatory framework. However, we failed to elicit substantial concerns about anything other than nanomaterials. This report, in particular this chapter and Chapter 3, therefore concentrates primarily on nanoscale materials. This focus leads naturally to an evaluation of the governance, regulatory structure and processes required to oversee their manufacture, use and disposal in Chapter 4. Whilst Chapter 4 again focuses primarily on nanomaterials, the general principles it sets out can be applied to all types of novel materials.

THE NANOSCALE

2.4   The small size of nanomaterials gives them specific or enhanced physico-chemical properties, compared with the same materials at the macroscale, which generate great interest in their potential for development for different uses and products.[2] Figure 2-I illustrates where the nanoscale fits into the wider spectrum of material dimensions.

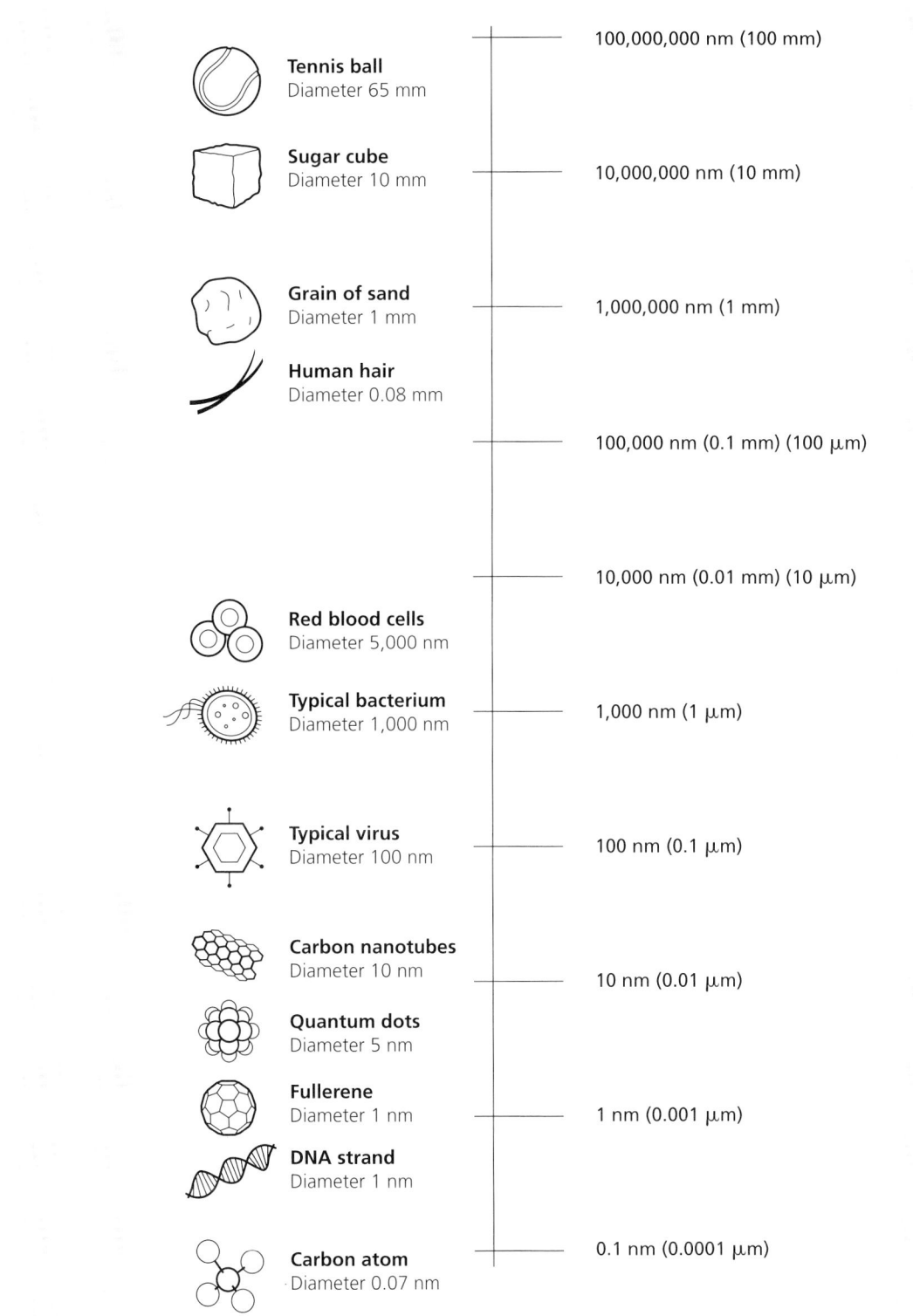

FIGURE 2-I
Length scale showing the nanometre in context

This diagram places the nanoscale in context. One nanometre (nm) is equal to one-billionth (1,000,000,000) of a metre, 10^{-9}m. Most structures of nanomaterials which are of interest are between 1 and 100 nm in one or more dimensions. For example, carbon Buckyballs (figure 2-III) are about 1 nm in diameter.

Tennis ball — Diameter 65 mm — 100,000,000 nm (100 mm)

Sugar cube — Diameter 10 mm — 10,000,000 nm (10 mm)

Grain of sand — Diameter 1 mm — 1,000,000 nm (1 mm)

Human hair — Diameter 0.08 mm

100,000 nm (0.1 mm) (100 μm)

10,000 nm (0.01 mm) (10 μm)

Red blood cells — Diameter 5,000 nm

Typical bacterium — Diameter 1,000 nm — 1,000 nm (1 μm)

Typical virus — Diameter 100 nm — 100 nm (0.1 μm)

Carbon nanotubes — Diameter 10 nm — 10 nm (0.01 μm)

Quantum dots — Diameter 5 nm

Fullerene — Diameter 1 nm — 1 nm (0.001 μm)

DNA strand — Diameter 1 nm

Carbon atom — Diameter 0.07 nm — 0.1 nm (0.0001 μm)

TERMS TO DESCRIBE NANOSCALE TECHNOLOGIES AND MATERIALS

2.5   Many terms are used to describe technologies and materials employed at the nanoscale, including 'nanoscience', 'nanotechnology', 'nanomaterials' and 'nanoparticles'. In evidence we have been told that it is difficult to point to a single definition that encapsulates 'nano'. Given the interdisciplinary nature of nanotechnology, however, a single definition is unhelpful and, as noted in Chapter 1, many believe that 'nanotechnology' as a term will cease to exist within the next decade because increasingly researchers and developers will select a material for its functionality, rather than for its size.[3] Nevertheless, a good working definition of a nanomaterial is one that is between 1 and 100 nm in at least one dimension and which exhibits novel properties.

2.6   Nanomaterials can have one, two or three dimensions in the nanoscale. One-dimensional nanomaterials include layers, multi-layers, thin films, platelets and surface coatings. They have been developed and used for decades, particularly in the electronics industry. Materials that are nanoscale in two dimensions include nanowires, nanofibres made from a variety of elements other than carbon, nanotubes and, a subset of this group, carbon nanotubes. Single-walled and multi-walled carbon nanotubes are two distinct types, but many variations within these two categories mean there are many nanotube types overall (figure 2-II). Their novel functionality affects their strength, electrical properties, thermal conductivity and ability to change properties with the addition of functional groups, meaning they have the potential to be used in a wide range of applications including composites, sensors and electronics. Nanowires are very fine wires, which can be made from a wide range of materials; they have applications in high-density data storage.

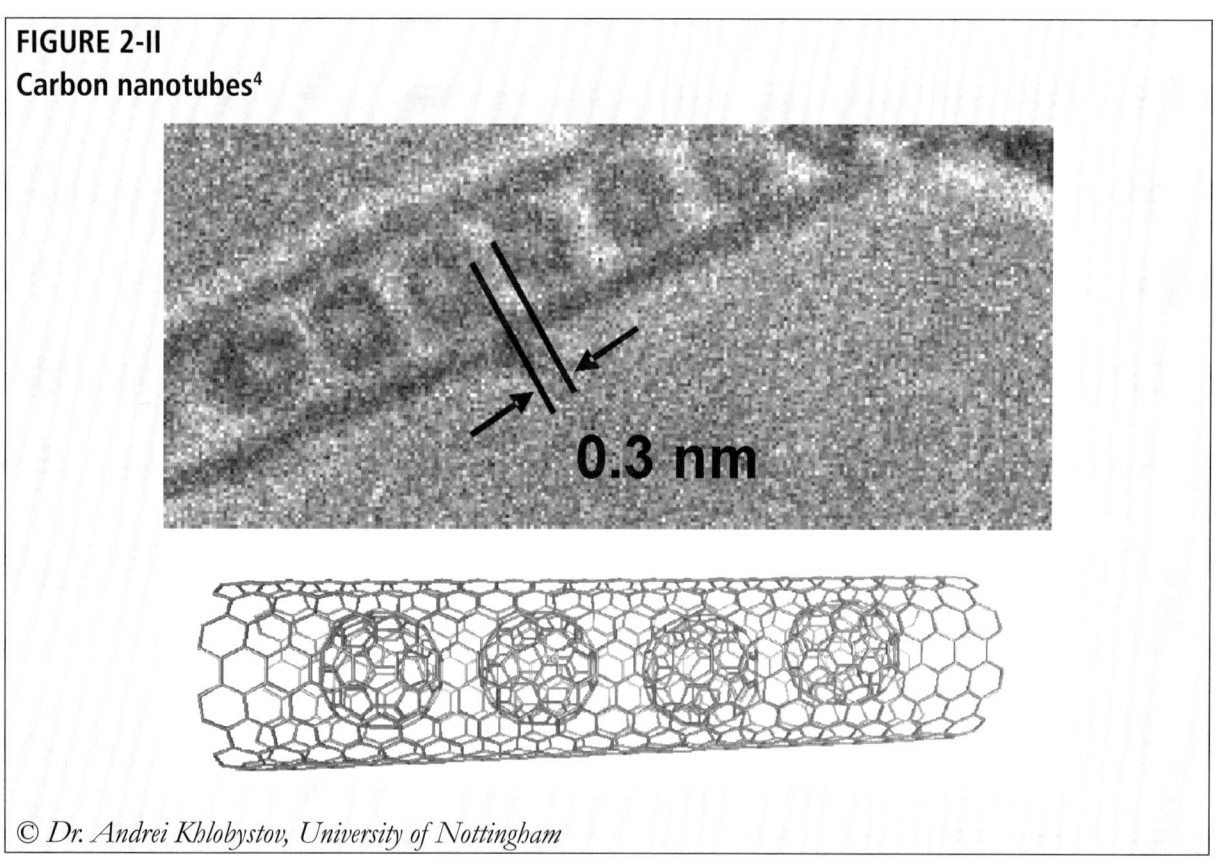

FIGURE 2-II
Carbon nanotubes[4]

0.3 nm

2.7   Materials that are nanoscale in three dimensions are known as nanoparticles and include precipitates, colloids and quantum dots (tiny particles of semiconductor materials). Nanocrystalline materials made up of nanometre-sized grains also fall into this category.[5] Nanoparticles exist naturally (for example, natural ammonium sulphate particles), but they can also be manufactured, as for example in the case of metal oxides such as titanium dioxide and zinc oxide. Metal oxide nanoparticles already have applications in cosmetics, textiles and paints and, in the longer term, could potentially be used for targeted drug delivery. Self-assembled nanoparticles and nanostructures are also being developed for use in targeted drug delivery. Dendrimers can include spherical polymeric molecules that are used in coatings and inks. Quantum dots have applications in solar cells and miniature solid state lasers.

2.8   Buckminsterfullerenes (also known as fullerenes and Buckyballs) are a class of nanomaterial of which carbon-60 (C_{60}) is perhaps the best known. C_{60} is a spherical molecule about 1 nm in diameter which comprises 60 carbon atoms arranged as the corners of 20 hexagons and 12 pentagons (figure 2-III). Potential applications include use as lubricants and electrical conductors.

FIGURE 2-III
C_{60} Buckminsterfullerene (also known as a Buckyball or fullerene)[6]

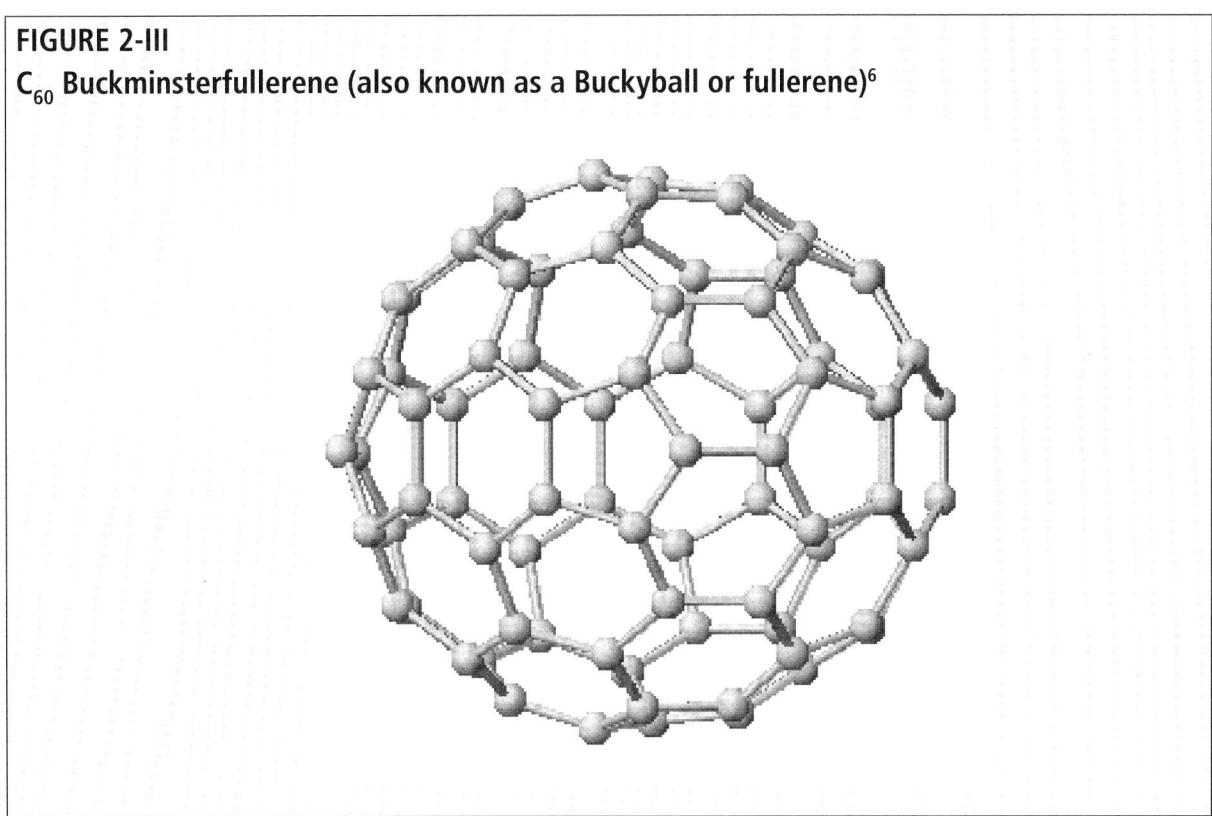

PROPERTIES OF MATERIALS AND NANOMATERIALS

2.9   As already noted (1.16 and 2.4), the properties and hence functionalities of nanomaterials can be very different from those of the bulk form and the component atoms and molecules. Furthermore, some properties being discovered have not previously been observed in traditional chemistry or materials science.[7] While the resulting difference in behaviour from the bulk form, or from the same material in the molecularly dispersed or atomic state, makes it possible to use nanomaterials in novel ways, it may also give rise to different mobility and toxicity in organisms and the environment.

2.10 The features of nanoparticles which underlie these properties and behaviour include: greatly increased surface area per unit mass; changes in the relative frequency of different component atoms at the surface (and hence in chemical reactivity); changes in surface charge; and modified electronic characteristics. The electronic features can become quantized, leading to so-called 'quantum effects' which can influence optical, electrical, magnetic and catalytic behaviour.[8] The strong surface forces and Brownian motion which may be exhibited at this size range are also important as they may play a significant role in the self assembly of nanostructures.

2.11 It follows that some novel properties of nanoparticles are predictable, but others will be unexpected compared with what is known from the existing science and technology base. Substances behaving in previously unobserved ways would fall into the 'revolutionary' category (according to the definition at 1.19). Examples include the catalytic properties of gold particles, the mechanical properties of carbon nanotubes and the optical properties of cadmium selenide quantum dots.

2.12 These effects and others described in more detail below are often well characterised in relation to the functionalities for which the new properties are being exploited. However, they are usually much less well characterised in terms of fate and behaviour in organisms and the environment, which may well present more demanding challenges.

2.13 While the basic principles employed in characterising substances for health and environmental effects are the same whether or not they are in the nanoform, certain properties are particularly or uniquely important in the case of nanomaterials. These include particle size, particle shape, surface properties, solubility, agglomeration and aggregation (appendix E). Furthermore, the way these properties determine behaviour can be profoundly influenced by extrinsic variables, such as temperature, pH, ionic strength of containing medium and presence or absence of light. In the following sections we illustrate the range of factors determining properties and functionalities. The challenges which this presents in relation to risk assessment and governance are discussed in detail in Chapters 3 and 4 respectively.

Composition

2.14 The composition of any material plays a central role in determining its properties, including reactivity, mobility and toxicity. A major advance being achieved with nanomaterials is to engineer composition more specifically to modify or enhance properties. In his seminal 1959 lecture *There's Plenty of Room at the Bottom*, the physicist Richard Feynman asked the pertinent question "What would be the properties of materials if we could really arrange the atoms the way we want them?".[9] The structural precision with which nanomaterials can now be engineered is providing the opportunity to address this question.

2.15 Composition can be further complicated by combining different substances to create a functional whole. Some nanomaterials are composites, consisting of a core (which is itself usually referred to as the nanomaterial) and a shell around the core produced either deliberately (as with many quantum dots) or unintentionally (as in the oxidation of zero-valent iron nanomaterials to form an iron oxide shell).[10] In addition, a surface active agent, sometimes called a capping agent, is often used in practical applications of nanomaterials. This is usually an organic molecule such as a polymer or surfactant. Small amounts of material (e.g. heavy metals), known as dopants, can also be added to alter the electrical and chemical properties of the nanomaterial.

2.16 All these aspects of composition are likely to affect behaviour in organisms or the environment. The polymer or surfactant layer, for example, is often used to impart colloidal stability and prevent aggregation and agglomeration. Nanomaterials with improved stabilising agents are being produced for specific applications at an increasing pace. Many are aimed at crossing biological membrane barriers to assist drug delivery and for other medical applications.[11, 12] However, because of this characteristic, these materials may be of particular concern if they enter the environment.

Size and shape

2.17 Size is one of the distinguishing characteristics of nanomaterials – their size range is such that size-dependent properties feature strongly in their behaviour. Prominent among such properties is surface area: table 2.1 shows how the surface area per unit mass increases significantly as size of particle decreases, a consequence of the increase in the number of particles. As many chemical reactions occur at surfaces, this means that nanomaterials may be relatively much more reactive than a similar mass of conventional materials in bulk form. This suggests that the weight thresholds embodied in legislation and regulation of chemicals and materials (e.g. the European REACH regulation, see Chapter 4) may not be valid for nanomaterials. The way surface properties affect reactivity is discussed further below (2.19).

2.18 At the nanoscale, shape may also be especially important, as experience with the needle-shaped asbestos fibres has shown. Nanomaterials exhibit a wide variety of shapes including particles, tubes, threads and sheets, as well as more ornate forms. For example, nanomaterials may be engineered as rods or dumb-bells.

TABLE 2.1
Influence of particle size on particle number and surface area for a given particle mass[13]

Consider a single particle the size of a basketball which is then broken into many smaller particles, each the size of a pea. Clearly the same mass of material can comprise one very large particle (the basketball) or thousands of smaller particles (the pea) but if one were to sum the total surface area of the smaller particles it would far exceed that of the larger particle. This table illustrates the phenomenon for an original large particle (diameter of 10,000 nm or 10 μm) broken down into smaller particles; by the time the constituent particles are 10 nm (or 0.01 μm) in diameter, it has produced 10^9 particles with an increase in surface area of a factor of 10^6.

Particle diameter (nm)	Relative number of particles	Relative surface area (as a factor)
10,000 nm	1	1
1,000 nm	10^3	10^2
100 nm	10^6	10^4
10 nm	10^9	10^6

Surface properties

2.19 Surface properties have a significant effect on how a material interacts with organisms and its behaviour in the environment. They change at the nanoscale; for example the forces binding individual surface atoms to the interior of a nanoparticle can decrease as the size decreases (and therefore the ratio of surface area to volume increases). This makes the surface atoms more reactive. Overall the surface chemistry of a substance will be influenced by the available surface area, the nature of the atoms at the surface, the charge at the surface and any surface modifications. Contaminants at the surface and structural defects can also modify properties. Surface chemistry could be a key indicator of the potential for harmful effects on health or the environment, although how the material is dispersed in the environment will also be an important factor. It will also influence how the material will attach to charged cells and biomolecules.

2.20 The charge at the surface influences how the substance will interact with other substances, for example in which solvents it will dissolve. Surface charge also affects whether particles will remain dispersed or will aggregate and agglomerate in any medium, which is important when considering how the material will be transported in the environment. In addition, surface charge together with other surface properties will affect the way in which a substance partitions between different phases, for example, how it will be sorbed. This has a major influence on bioavailability, mobility in the environment and penetration to sites of toxic action in organisms.

2.21 Surface chemistry can be markedly affected by defects, dopants or impurities, adding considerably to the complexity of factors that need to be taken into account when considering the surface activity of a material.

Solubility

2.22 A key factor determining the impact of a nanomaterial on the environment is how it is dispersed. Materials which are freely soluble in water generally move readily through aqueous environments, whereas insoluble particles have different transport mechanisms. A further complication is that particles which are discrete under laboratory conditions may aggregate in aqueous systems in the environment (2.23) with consequent effects on transport pathways. It is therefore important to consider the different ways in which nanomaterials may be dispersed in other media. These are described in appendix F.

Aggregation

2.23 While discussion of nanoparticles tends to focus on discrete particles, in practice particles often aggregate (i.e. adhere together), significantly changing behaviour, for example partition and transport in the environment. To prevent aggregation, the surface of the particle can be modified or it can be suspended in a medium that limits aggregation – a well-established aspect of colloid science.

2.24 The potential for particles to aggregate is determined by the repulsive force that particles experience when suspended in a specific medium. The lower this is, the less the electrostatic force of repulsion between adjacent particles, which increases the likelihood of them coalescing

16

to form a larger entity. The environmental behaviour of aggregated and single particles will differ, with the larger particles tending to settle in the medium and smaller particles 'going with the flow'.[14]

APPLICATIONS AND USES OF NOVEL MATERIALS

EXAMPLES OF NANOMATERIALS AND THEIR USES

2.25   The introduction to this chapter (2.6-2.8) only begins to illustrate the diversity of nanomaterials. They do not share a common scientific basis or technology, nor do they fit into a single group of products or markets which share one common feature.[15] Their specialised properties described above, and hence functionalities, mean that nanotechnologies and nanomaterials have the potential to be developed and used widely through nearly all sectors of life, including communications, health, housing, energy, food and transport (1.5-1.7).[16, 17] Table 2.2 shows examples of nanomaterial products used in the automotive industry. Box 2A describes the application of nanotechnology to medicine. In all these applications, nanomaterials are being exquisitely designed for very specific purposes.

TABLE 2.2
Examples of nanomaterial products used in the automotive industry[18]

Product	Nanomaterial	Function/use
Carbon black	carbon nanoparticles	Improves mechanical properties of car tyres
Ceramiclear	ceramic nanoparticles	Scratch resistant clear coatings for vehicles
Components for fuel line and tank	carbon nanotubes (composites)	Anti-static agents
Carbon nanotube polymer composite	carbon nanotubes	Allows electrostatic coating
Nano-TPO	nanoclay thermoplastic composite for exterior parts	Improves mechanical properties
Schott Conturan®	glass nanocoatings	Anti-reflection coating for speed indicator glazing
OnStar Mirror	functional nanolayer	Auto-dimming mirrors
Catalyst materials	rare earth and platinum group metal nanomaterials	Catalytic converters

17

BOX 2A	NANOMEDICINES

The application of nanotechnology to medicine results in a whole new class of products known as nanomedicines. Their application ranges from use in diagnostic imaging[19] to use as scaffolds for tissue regeneration in orthopaedic implants.[20] Intelligent nanomaterials are also being designed as biosensors.[21] However, their widest use has been as drug delivery systems.[22]

To provide effective drug delivery, passive targeting with particular types of nanoparticle exploits vascular differences between the target tissues, e.g. between cancer cells and normal tissue, whereas active targeting is achieved by linking the polymer that comprises the nanoparticle to molecules such as monoclonal antibodies that specifically recognise cell surface receptors of interest.[23] Nanopolymers preferentially access tumours because they have larger pores (up to 2,000 nm in diameter) in their capillaries, compared to healthy tissues. The liver also has larger (100-200 nm) than normal pores explaining the increased uptake of nanoparticles by this organ.

Most frequently the nanomedicine is made up of an outer shell of a hydrophilic polymer (e.g. polyethylene glycol) and an inner core of hydrophobic polymer (e.g. polyaspartate) to generate composites ranging from 12-85 nm. An alternative structure is the dendrimer, which is a repeatedly branched polymer containing cascades of branches with a core surrounded by a shell.[24] By incorporating toxic anti-cancer drugs into the core, preferential uptake and prolonged drug release into a tumour occurs with less systemic toxicity.[25] Incorporation of anti-cancer drugs into nanomaterials also prolongs their effective life in lymphatic tissue inhibiting tumour spread (metastases) to these sites.

Application of these principles in the field of nanomedicine is also allowing nanomaterials to be used in neural regeneration and neuroprotection, as well as targeted drug delivery across the blood–brain barrier, which may be of special relevance in the treatment of neurodegenerative diseases.[26]

2.26 The nanomaterials market is growing rapidly. The Woodrow Wilson Center's database lists over 600 products self-identified as containing nanomaterials currently available in the global marketplace.[27] The products of nanotechnology can be found in paints, fuel cells, batteries, fuel additives, catalysts, transistors, lasers and lighting, lubricants, integrated circuitry, medical implants, water purifying agents, self-cleaning windows, sunscreens and cosmetics, explosives, disinfectants, abrasives and food additives.[28]

2.27 Nanosilver, various forms of carbon, zinc oxide, titanium dioxide and iron oxide make up the majority of nanomaterials in use, although others, for example nanogold, have started to enter the market.[29] The worldwide market for carbon nanotubes is currently $700 million, and expected to grow to at least $3.6 billion.[30] For titanium dioxide it is estimated at $314 million (5,000 tonnes), expected to grow to $471 million in the long term. The market for zinc oxide is estimated at $0.79 millions (18 tonnes). Common nanomaterials such as carbon black ($8 billion) and nanosilica ($3.14 billion) will have lower growth.[31] Overall the nanomaterials market is estimated to be worth about $30 billion per year.[32]

2.28 The growth in nanotechnology is also illustrated by the number of patents taken out on nanomaterials. Figure 2-IV shows the number of patents registered globally from 1990-2006 with any of the following in their titles: nanoparticle, nanorod, nanowire, nanocrystal, nanotube or carbon nanotubes. The acceleration in patenting is remarkable – more than a doubling in number of patents every 2 years.

FIGURE 2-IV
Trends of patents on nanomaterials (1990-2006)[33]

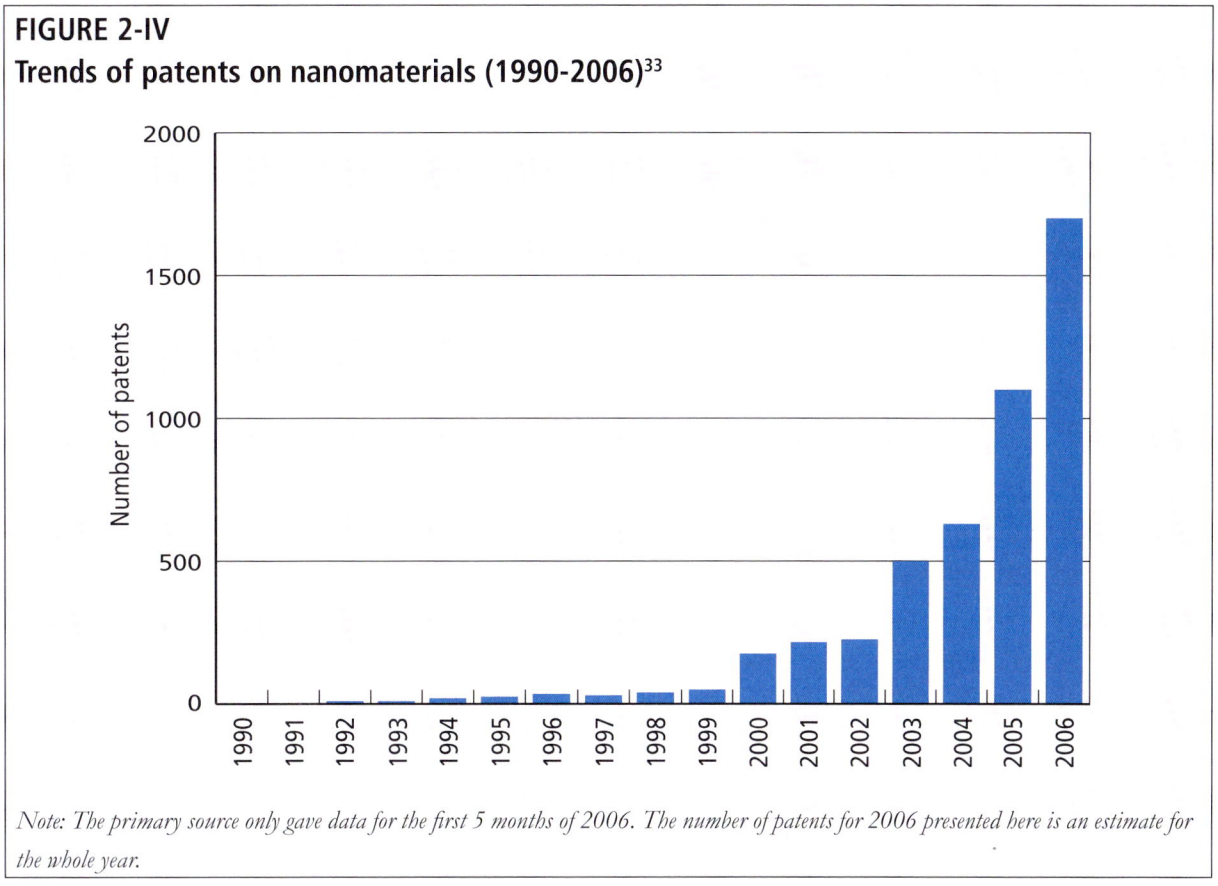

Note: The primary source only gave data for the first 5 months of 2006. The number of patents for 2006 presented here is an estimate for the whole year.

2.29 There is also much interest in developing nanomaterials to benefit the environment, for example in the following fields:

- energy generation and storage, including electricity storage, fuel cells, and hydrogen storage and generation;

- water, air and land quality, including environmental sensors, soil remediation, agricultural pollution reduction and water purification; and

- energy efficiency, including insulation, lighting, engine and fuel efficiency, 'lightweighting' of materials and the development of other novel materials with environmental benefits (e.g. the development of ultra hydrophobic coatings to reduce the icing-up of wind turbine blades).[34]

2.30 It is clear that there is a great deal of interest in novel materials, particularly nanomaterials, and they promise much in the way of benefits to society and the environment. There remain concerns however that the potential benefits of some applications of nanotechnology have been exaggerated, in that a benefit discovered under laboratory conditions may not be realised on a commercial scale. Equally, concerns have been expressed about the potential environmental and human health impacts of these materials and technologies (Chapter 3). The benefits perceived and levels of concern shown by different individuals and organisations depend strongly on, and indeed reflect, the different 'world views' summarised in Chapter 1. Some see only benefits, others only problems. The aim of this report is to steer a course that will allow the benefits of nanomaterials to be realised and their possible risks to be avoided. This is challenging in the face of very great uncertainties and a profound lack of detailed information about both possible benefits and possible risks in a rapidly evolving field. Central to our approach is the recognition that it is the properties and functionalities of novel materials that matter, not how they are made, nor simply their size (Chapter 1).

2.31   Some commentators have envisaged four generations of nanotechnology products and processes with potential for development between 2000 and 2020[35] (figure 2-V), though some claims need to be treated with caution since the timescales for change are inherently difficult to predict. It is clear, however, that even first and second generation nanotechnologies (which already exist) present major challenges in terms of understanding their social and environmental implications. The implications of what are seen as third and fourth generation nanotechnologies are profound and represent a significant step change in the challenges to the regulatory system and to the need for societal engagement.

FIGURE 2-V
Four generations of products and processes[36]

The first generation comprises passive nanostructures, which are already being produced and made available on the commercial market, through to molecular nanosystems which are thought to include so-called 'designer molecules'. Whilst we broadly accept the classification into four generations of product types, we are sceptical about the timescales proposed. Third and fourth generation products may take much longer to reach the marketplace than suggested here.

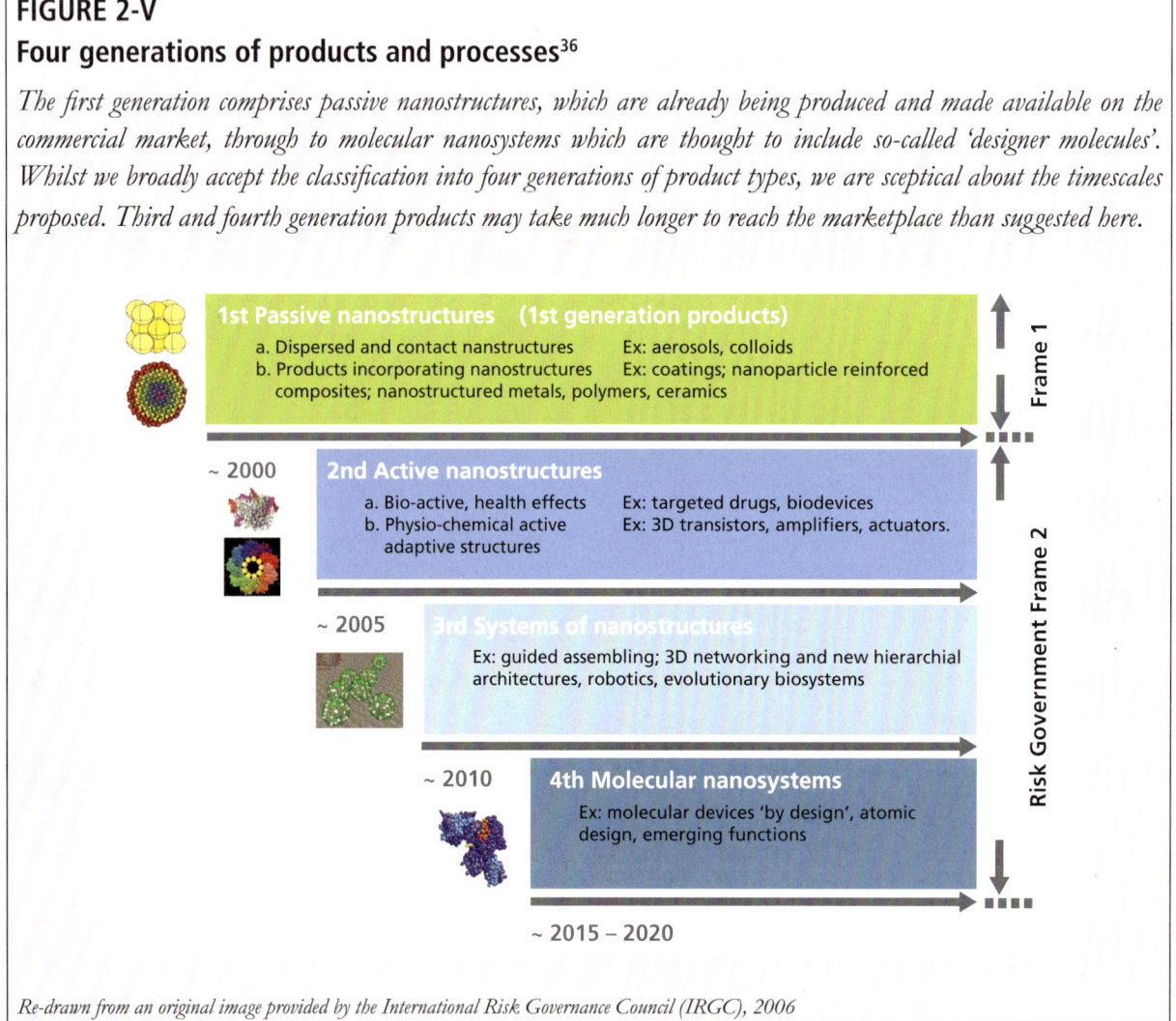

Re-drawn from an original image provided by the International Risk Governance Council (IRGC), 2006

THE NANOTECHNOLOGY INNOVATION SYSTEM

2.32   When considering the potential impacts of nanomaterials and how to manage them, it is not enough just to consider their properties in isolation. The development of nanomaterials and the production methods used to manufacture them for a particular application also need to be taken into account.

2.33   Innovation systems analysis is one of the tools available to help develop policy for managing nanotechnology. The empirical description of an innovation system requires the identification of the main actors, their linkages and the rules or institutions that govern their behaviour (regulatory regimes, intellectual property rights, etc.). The nanomaterials innovation system is international, heterogeneous and complex. It involves a wide range of actors and processes, including academic research, which is supported by specialised infrastructure and is mainly publicly funded. Academic research interacts with small suppliers and large manufacturers which provide nanomaterials for user firms in a variety of sectors. More traditional and less technology-intensive nanomaterials may be provided by firms specialised in manufacturing. User firms incorporate nanomaterials into their products for consumers. A key driver for innovation is consumer desire for new functionalities, which can be delivered (although not always) by nanomaterials. National governments and the European Union promote research via funding bodies, with both generic funding and specific funding, in particular for infrastructure. They also regulate nanomaterials via regulatory agencies which are co-ordinated through international regulatory initiatives and advisory groups (e.g. the Organisation for Economic Co-operation and Development (OECD)). Other actors include learned societies (e.g. the Royal Society) or industrial organisations (such as the Nanotechnology Industries Association).

2.34   A key issue in understanding the system of innovation in nanomaterials is that the great majority of nanomaterials are not consumer products to be sold to an end user, but 'capital' materials to be used by other industries in order to make new products. In this sense most nanomaterials can be understood as 'products for process innovation'. This is why supplier and manufacturing firms occupy the central position in nanomaterials innovation systems. The innovation system for nanomaterials can therefore be conceptualised as an 'hourglass model' (figure 2-VI) in which a variety of scientific disciplines support the development of a number of technologies for the fabrication of nanomaterials, which then serve many different economic sectors.

2.35   In the intermediary role of supplier and/or manufacturer of nanomaterials, three types of firm can be distinguished: small suppliers; specialised manufacturers; and larger suppliers/ manufacturers. Small, new firms emerge as specialised suppliers in particular niches created by radical innovations (often stemming from academia). Such high-performance products are unlikely to have large markets, but can produce substantial improvements in the operation of other technical systems, for example in medicine or biotechnology. Firms producing them might therefore be expected to work closely with their customers to co-develop products for very specialised submarkets.

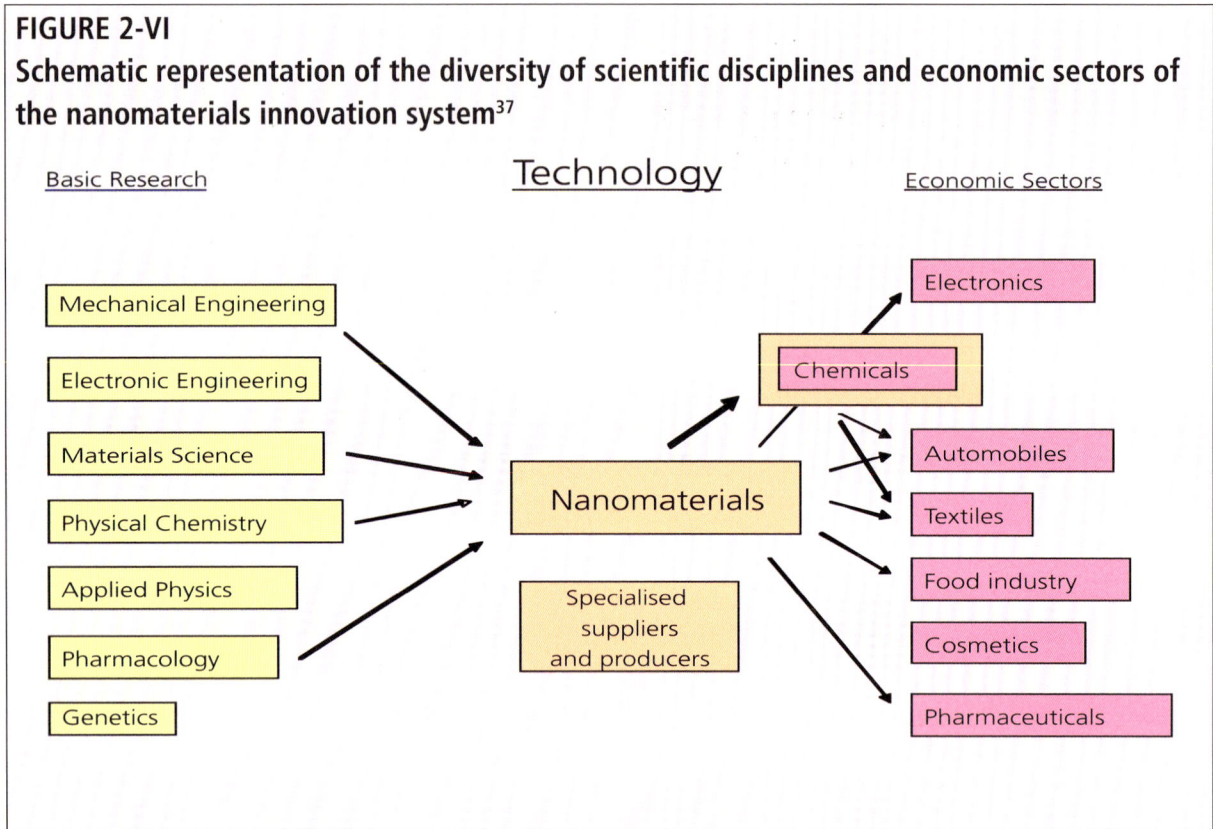

FIGURE 2-VI

Schematic representation of the diversity of scientific disciplines and economic sectors of the nanomaterials innovation system[37]

2.36   The second type of actor is the small specialised manufacturer with expertise in the large-scale production of certain products. In general, this large-scale production relies less on academic knowledge and more on in-house industrial know-how for process innovation.

2.37   Thirdly, in the case of science-based sectors such as chemistry or electronics, the R&D and production capabilities of the large corporations allow them to be active both in the development of new nanomaterials and in their production. Thus, publication and patent analysis shows that transnational corporations dominate the rankings.[38] In spite of this evidence of research activity, some of these large firms appear to be less eager than small supplier firms in the branding of their materials as nanotechnology. This seems particularly the case in nanoelectronics, where one- and two-dimensional nanostructures are now common. Finally, in supplier-dominated sectors, such as textiles, innovation is carried out by the specialised suppliers. For example, innovation in textiles will be provided by the producers of fibres, and some of these will be supplied by producers of nanomaterials.

2.38   The co-existence and mutual interdependence of various types of firms results from the different degree to which innovation in nanomaterials is a radical shift from previous materials technologies. In general, at the initial stages of an innovation, customers demand, and are prepared to pay for, high performance. Such innovation focuses predominantly on product (in this case a 'product for process') rather than process innovation. Hence, for radically new nanomaterials such as carbon nanotubes there are now many small technology start-up firms, often related to academia and supplying some niche markets.

2.39 However, it is expected that, due to the cumulative nature of knowledge and the need for expensive and complex negotiations with customers and regulators, only larger firms will be able to afford the long-term investments needed to develop research, together with large-scale production and commercialisation.[39] This is because large firms have an advantage in finding markets for well-characterised materials in search of an application; and, second, improvements in performance come from cumulative experiments, typically conducted in R&D laboratories, which build on links to public science. In spite of the foreseeable dominance of large firms, there appears to be space for specialised nanomaterials manufacturers following established production techniques. These can continue to produce nanomaterials building on their previous expertise and do not require a new knowledge base or very accurate characterisation.

2.40 The innovation system for nanomaterials ranges from large multinational companies to small, often highly innovative, high-tech firms. The typical innovation processes would be expected to involve close connections with the science base and regulators, and links between suppliers and users along supply chains. The nanomaterials sector also covers a variety of different technologies and is global in its coverage, drawing on international knowledge and generating products that are manufactured and sold beyond the boundaries of a single nation state. Such diversity means that it is unlikely that the entire sector can be regulated satisfactorily by a single regulatory body. This issue is explored further in Chapter 4.

PATHWAYS AND FATE OF NANOMATERIALS IN THE ENVIRONMENT

2.41 The potential of nanomaterials to be used across a wide range of sectors means that the number of possible routes for nanomaterials to reach and enter organisms and the environment is high. Nanomaterials enter ecosystems from both point and diffuse pollution sources.[40] They may be discharged directly into rivers or the atmosphere by industry, or inadvertently escape as products, such as paints, cosmetics, sunscreens and pharmaceuticals, are used or disposed of in the environment.

2.42 In view of the apparent absence of evidence of harmful impacts of manufactured nanomaterials in 'real world' situations, we can only examine the plausibility of damage based on the extrapolation of evidence from laboratory investigations and occupational exposure studies on dust and other substances. As is often the case in toxicology, the approach which we are left with is to identify the characteristics of the manufactured nanomaterial in question, determine its bioavailability and persistence in natural settings, then use data derived from measured concentrations in the environment as well as toxicological research in the laboratory to assess hazards and risks.

2.43 There is a widespread consensus that comprehensive characterisation of nanomaterials, both before and during exposure, is required to understand fully their potential fate and effects.[41] Such characterisation is lacking in the vast majority of studies. Even under controlled laboratory conditions, the true size distribution of nanoparticles may differ significantly from the advertised sizes of commercially-supplied materials.[42] Sample preparation and conditions of analytical quantification may alter sample integrity, so that qualitative and quantitative analyses do not adequately describe exposure conditions in environmental matrices.[43, 44]

2.44 Despite this, it is notable that many nanomaterial manufacturers, including those engaged in the production of nanomedicines and food additives, appear to feel confident that their products pose little or no threat to human health or the environment. Most studies to date report chemical composition, but few refer to the size distribution or surface charge properties of nanoparticles, or to the size and shape of aggregates that form at higher concentrations in aqueous media.[45]

2.45   Once nanomaterials are released into the environment a variety of processes can modify their functional properties and influence the likelihood of their uptake into living organisms. Some of the key properties have already been outlined above (2.9-2.24). The fate of nanomaterials in aquatic ecosystems depends largely on their solubility in the aqueous phase and their potential for aggregation.[46] The aggregation behaviour of nanomaterials is especially important. Aggregate size, morphology and kinetics alter with nanomaterial type and other environmental factors. Aggregation processes clearly influence the environmental fate and behaviour of nanomaterials, and the concentration to which organisms are exposed. However, the potential reduction in biological effects with aggregation should not be over-emphasised. For example, aggregates passing into natural waters may undergo re-suspension, disaggregation and other processes prior to incorporation within sediments and permanent loss.[47] There is evidence that large, discrete particles may be considerably less toxic than similarly-sized aggregates of nanomaterials with the same chemistry. Thus, the novel properties of nanomaterials may persist even when in the aggregated form.

2.46   The release of carbon nanotubes, nanoparticles of zero-valent iron, titanium dioxide and fullerenes into water can result in their aggregation.[48] Both the extent of aggregation and the size range of the aggregates vary with particle character and environmental conditions (2.13). Particles tend to aggregate in saline conditions[49] and may adsorb to sediment, algae, soil or to the surface of gills and epithelial cells, shells and cuticles.[50] Even very small changes in salinity can lead to changes in colloid formation and aggregation,[51] with the prediction that colloidal manufactured nanoparticles will precipitate from solution when moving from fresh to estuarine environments.[52]

2.47   In contrast to the aqueous environment, once in soils, manufactured nanomaterials may be temporarily fixed or be degraded through chemical, photochemical or microbial processes.[53] Some nanomaterials are strongly sorbed to soil particles,[54] whilst others remain relatively mobile,[55] depending on particle size and physico-chemical characteristics.

2.48   Recent reviews have addressed the issue of nanomaterials in the environment.[56] A major area of debate is whether the physico-chemical properties of nanoparticles and nanotubes can be related to their potential environmental fate and toxicological effects. The consensus at present is that we are unable to make this connection, but that with further research it might be possible.[57] We wonder whether a profitable approach might be the application of a modified form of the QSAR (quantitative structure–activity relationship) methodology which has been successfully used to assess the toxicity of a vast range of organic chemicals. Factors such as shape are likely to be more important for nanoparticles than for conventional chemicals, where molecular properties are the basis of QSARs.

2.49   However, we note that in some fields of research the characteristics of particular nanoparticles are apparently already well understood. Nanomedicines provide an example. For instance, nanotubes have been chosen for the delivery of drugs to highly specific cellular target sites using knowledge of the physico-chemical properties of the nanotubes themselves (box 2A).

2.50   A key challenge for ecotoxicologists is to test the toxicity of the nanomaterial itself and eliminate the confounding effects of vehicle solvents or uncharacteristic solvent-induced effects in the nanomaterial. We discuss these problems further in Chapter 3.

2.51   Investigation of the specific links between physical chemistry, bioavailability and subsequent effects will help to reveal whether exposure to nanomaterials is likely to be significant in the environment. For example, ecotoxicological studies on organisms within the water column may be less relevant if the nanoparticle of interest aggregates rapidly and completely. If this is the case then benthic organisms (e.g. deposit-feeding molluscs and annelid worms) are more likely to be environmentally relevant target organisms than free-swimming pelagic species (e.g. fish and water fleas).

2.52   Whatever the current level of exposure of organisms, the increasingly widespread use of different kinds of nanoparticles and nanotubes, and the predicted exponential increases in production volumes, will undoubtedly lead to greater exposure of biota within all environmental compartments in the future.[58] Industrial products and wastes tend to end up in streams, rivers and estuaries and are ultimately discharged to the sea. Physical and chemical processes in each compartment are likely to alter the properties of nanomaterials, e.g. UV exposure can alter the coatings of fullerenes[59] and quantum dots,[60] making risk assessment throughout the nanomaterials life cycle more difficult.[61]

THE ENVIRONMENTAL LIFE CYCLE OF NANOMATERIALS

2.53   The life cycle of a manufactured product includes all the processes and activities that occur from initial extraction of the material (or its precursors) from the earth to the point at which any of the material's residuals are returned to the environment. A diagram of a typical life cycle is shown in figure 2-VII.

2.54   Life cycle assessment can be used to assess material and energy flows throughout the life cycle of a given product or process. It can be used to identify environmental impacts, inefficient processes, high energy use and exchanges of materials with the environment. Life cycle assessment is a useful means of identifying the different actors that are involved, as well as the linkages between them. It is important to consider the whole life cycle of a nanomaterial when looking at its potential impacts on the environment and human health, as its properties can change over time or throughout different stages of its life cycle. For example, the consideration of post-use options, such as recycling or disposal, is vital because once the product or material has served its intended purpose it will enter either a new system (through recycling) or the environment (through the waste management system). Concerns about the possible impacts of nanomaterials on the environment and human health during their life cycle are explored further in Chapter 3.

2.55   Failure to consider the full life cycle of a manufactured nanoparticle (or any other novel material) can lead to serious errors in judgement about benefits. For instance, a particular material may greatly enhance the performance of an energy storage device, but if its manufacture or disposal leads to environmental contamination or human health risks, the benefits may be outweighed by the disadvantages. During research for this study, we found a worrying incidence of very myopic views of 'benefits' because a full life cycle assessment of the material had not been considered by its proponents.

FIGURE 2-VII
A representation of a typical life cycle for manufactured products[62]

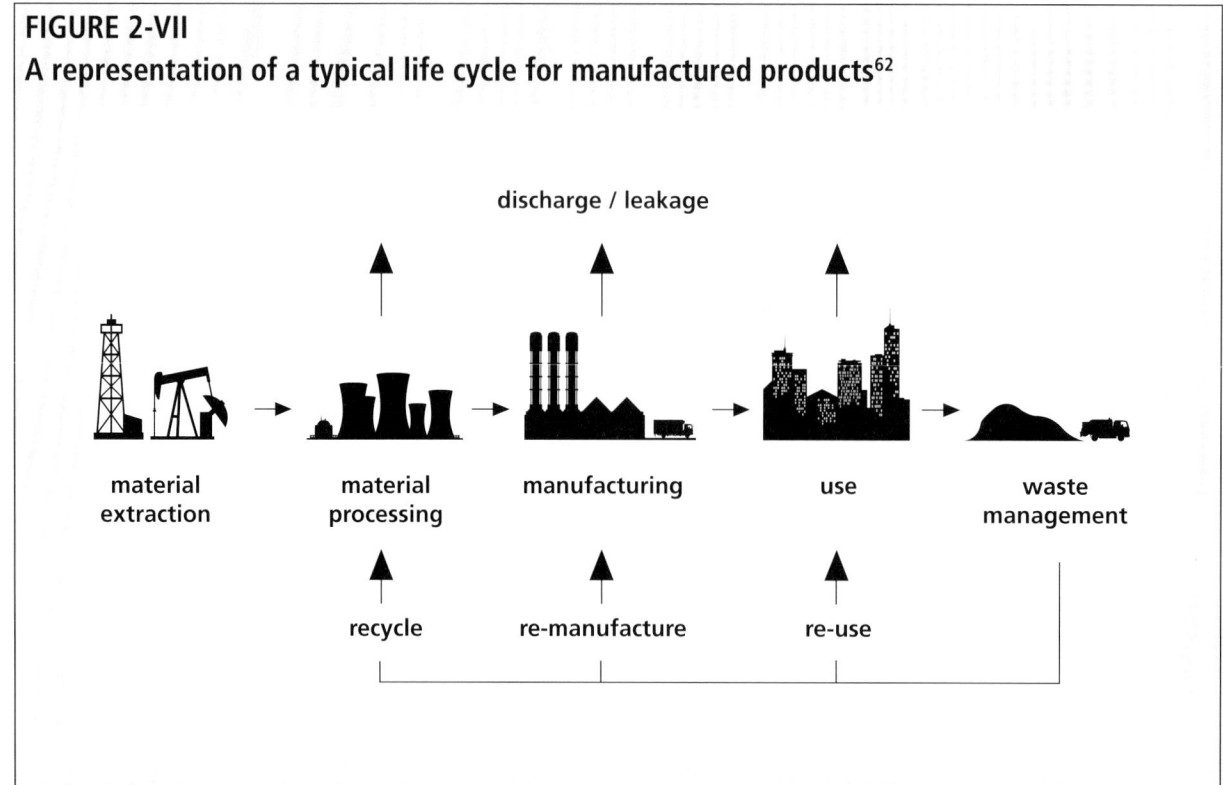

CONCLUSIONS

2.56 In this chapter we have reviewed the properties, purpose and production of nanomaterials, showing how their unique properties and behaviour make them potentially useful in a wide range of products and applications but which at the same time may mean that they have novel effects on organisms and the environment. In turn, this means that there are many potential routes for nanomaterials to enter ecosystems throughout their life cycle. We have also described the innovation system for nanomaterials, looking at what its features might mean for their governance. In the next chapter, we consider how the properties of nanomaterials affect their behaviour in organisms (including humans) and the environment during all stages of their life cycle.

Chapter 3

ENVIRONMENTAL AND HEALTH IMPACTS OF MANUFACTURED NANOMATERIALS

INTRODUCTION

3.1   In the previous chapter we reviewed the physico-chemical properties of nanomaterials and briefly examined the ways in which they might be released into and distributed in the environment. In this chapter we consider the possible interactions between nanomaterials and living organisms. The evidence available is from studies performed in the laboratory with animals, plants, micro-organisms, fungi and various cell lines. It suggests that there is a plausible basis for concern that harmful effects might arise. At present, however, we have not seen evidence of actual ecological damage or harm to humans resulting from exposure to manufactured nanomaterials.

3.2   We move on to consider the value and relevance of current risk assessment procedures and how they might be improved, and examine the feasibility of performing toxicological evaluations over an appropriate timescale. Finally, we evaluate the need for environmental monitoring and surveillance to detect unexpected effects. Before we start however, it is worth reflecting on the research effort that has so far been undertaken.

3.3   Concern about the potential harm associated with manufactured nanomaterials has stimulated much research activity in the UK and beyond. Within Europe as a whole, the European Commission relies on the considered opinion from three independent non-food scientific committees when formulating policy proposals on public health and the environment. These are the Scientific Committee on Consumer Products, the Scientific Committee on Health and Environmental Risks and the Scientific Committee on Emerging and Newly Identified Health Risks (SCENIHR). The European Commission also receives advice from the European Food Safety Authority, the European Medicines Evaluation Agency, the European Centre for Disease Prevention and Control, and the European Chemicals Agency.

3.4   In 2005 the European Commission approached SCENIHR to request its opinion on the appropriateness of existing risk assessment methodologies (as described in the Technical Guidance Documents (TGDs) of the chemicals legislation) for application to nanomaterials.

3.5   In its response, SCENIHR acknowledged that not all nanoparticle formulations would induce more pronounced toxicity than their bulk form, and therefore, would have different toxicological properties.[1] Consequently, their risks should be assessed on a case-by-case basis. It found that the TGDs made little reference to materials in a particulate form but that the methodologies they described were likely to identify potential hazards associated with nanoparticles. However, the Committee warned that the standard metric of mass concentration used to express the nanoparticle dose used in the development of dose–response relationships may require further attention with particle number, concentration and surface area perhaps being more appropriate (table 2.1). The fate and effects of nanomaterials in the environment are not well understood and the characteristics of various nanoparticles under different environmental conditions need to be determined. SCENIHR further suggested a number of potential improvements to current risk assessment methodologies to account for nanomaterials. These included:

- investigating nanomaterial characteristics under a range of environmental conditions;

- measuring the agglomeration and disagglomeration of nanomaterials under various environmental conditions;

- developing test methods for nanoparticle translocation, cellular uptake and toxicity mechanisms;

- defining a set of reference materials for nanoparticle testing;

- the development of QSARs (quantitative structure–activity relationships); and

- the development of validated tests for nanomaterial mutagenicity, genotoxicity and carcinogenicity.

3.6 Finally, SCENIHR recommended that a tiered approach to testing should be developed in order to produce a framework for assessing the potential risks associated with nanomaterials.

3.7 At a wider international level, the Organisation for Economic Co-operation and Development (OECD) first acknowledged concern about the safety of nanomaterials in its Chemical Committee in November 2004 and has subsequently made efforts to co-ordinate international research in this area. Subsequent sessions and workshops eventually led to the establishment of the OECD Working Party on Manufactured Nanomaterials (WPMN) in 2006. Its chief objective is to "promote international co-operation in human health and environmental safety-related aspects of manufactured nanomaterials (MN) in order to assist in the development of rigorous safety evaluation of nanomaterials".

3.8 As part of a wide-ranging programme of research, the WPMN established a project entitled *Safety Testing of a Representative Set of Manufactured Nanomaterials*, the objective of which was to agree and test a representative set of manufactured nanomaterials using appropriate test methodologies. The first stage in this project was to agree which nanomaterials would form the priority list of candidate 'representative' nanomaterials. The use of the phrase 'representative set' was taken to include nanomaterials already in or close to commercial use. It was also intended that the list would form a group of exemplar reference materials to support the measurement, toxicology and risk assessment of nanomaterials. The priority list was not considered to be definitive, but rather to act as a time-dependent indicator of materials considered to be important at any one time. The list could therefore change with time as new nanomaterials were developed. The 14 nanomaterials chosen to form the initial priority list for testing are as follows:

- fullerenes (C_{60});

- single-walled carbon nanotubes;

- multi-walled carbon nanotubes;

- silver nanoparticles;

- iron nanoparticles;

- carbon black;

- titanium dioxide;

- aluminium oxide;

- cerium oxide;

- zinc oxide;

- silicon dioxide;

- polystyrene;

- dendrimers;

- nanoclays.

3.9   Having formulated the test list of nanomaterials, the second stage of the project sought to formulate an understanding of their intrinsic properties relevant for the assessment of exposure and effects regarding human and environmental health. A list of endpoints was drawn up which included details necessary for physico-chemical properties and material characterisation, environmental fate, environmental and mammalian toxicology, and material safety to be assessed. The WPMN then launched its 'sponsorship programme' requiring different countries to sponsor the testing of specific nanomaterials.

3.10 In recent years in the European Union (EU) considerable effort has been channelled into a substantial programme of research and series of workshops by DG SANCO (the Directorate General for Health and Consumer Affairs) and other Directorates General in the European Commission. While we highly commend the efforts of the OECD and the European Commission and believe their approach to be very necessary, we are left with the feeling that the task in hand is formidable and that the time required to achieve an acceptable risk assessment methodology very short. Our reasons for adopting this stance are explained in this chapter. It is evident that the development of products containing nanomaterials has been much faster than any corresponding collection of environmental health data. Consequently the ability of regulatory bodies to incorporate this information into their policy thinking has been severely hampered. This is illustrated in figure 3-I, which shows the time lag between the emergence of products containing nanomaterials and the development of any associated environmental health information, and the subsequent lag in bringing this information to bear in policy considerations.

FIGURE 3-I
The emergence of information[2]

Schematic representation of the gap between the emergence of products containing nanomaterials in comparison to the generation of environmental health and safety data (EHS) and their subsequent use by regulatory agencies. The diagram is purely qualitative.

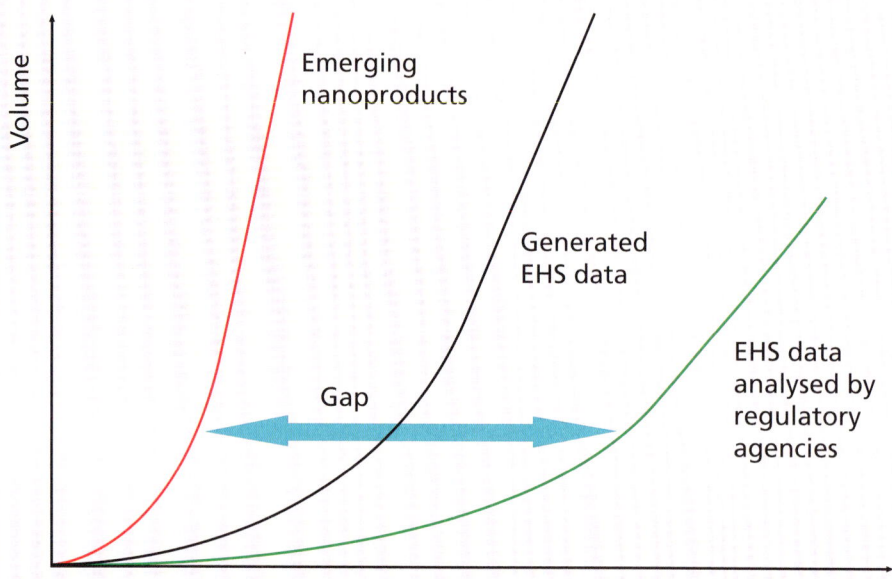

Reproduced by kind permission of Dr I. Linkov, US Army Engineers Research and Development Centre, Brookline, MA.

3.11   Free manufactured nanoparticles and nanotubes are likely to present the most immediate toxicological hazard to living organisms as they are at liberty to interact with organisms in the wider environment.[3] There is not the same level of concern regarding fixed nanomaterials, although there is clearly potential for them to become detached and enter natural ecosystems, especially when products containing them abrade or weather during use or when they are disposed of as waste or recycled.[4] Broken fragments of objects with intact surface coatings of nanomaterials provide an example of how fixed nanoparticles might pose a threat if they enter the environment.

3.12   Evidence presented to us has often been contradictory. On the one hand some environmental scientists and policy-makers feel strongly that the threat posed by most nanomaterials is small, whereas others are clearly worried about the possible toxicity of some nanomaterials, both to the wider environment and to human health. In particular, concern was expressed about an increased risk of lung and cardiovascular damage in humans, and effects on microbial communities and sediment-feeding organisms in natural ecosystems exposed to nanomaterials. There is a consensus that mechanisms of toxicity are poorly understood and that, with minor exceptions,[5] appropriate ecological studies have not been undertaken, including studies that address food chain transfer and multi-generational effects.[6] Currently it is extremely difficult to evaluate how safe or how dangerous nanomaterials are because of our complete ignorance about so many aspects of their fate and toxicology.

3.13 We described in Chapter 2 how manufactured nanomaterials, such as carbon nanotubes, have been produced in many different forms with a wide range of properties. Moreover, in each environmental compartment the various forms can be bioavailable to very different extents and exert very different toxicological properties.[7]

3.14 Little attention has been paid to the potential effects of nanoparticles generated through attrition of man-made products. For example, nanoparticles are produced through wear and tear on tyres and brake linings, and from clothes containing nanofibres that may be abraded during wearing and washing.

3.15 Microscale and, almost certainly, nanoscale fragments of plastic in sediment resulting from the gradual physical breakdown of plastic items (plastic bags, bottles, cigarette lighters, etc.) have been reported in the marine environment.[8] Sediment-feeding organisms, such as snails and worms, ingest the plastic particles and may be damaged either by the particles themselves or by pollutant chemicals bound to their surfaces.

ENVIRONMENTAL BENEFITS OF NANOMATERIALS

3.16 As well as the potential threats posed to ecosystems and humans, we have been alerted to the potentially wide range of benefits to the environment and human health that might accrue from the use of nanomaterials with their new or enhanced properties (see Chapter 2).

3.17 In some countries nanomaterials have been deliberately introduced to improve degraded ecosystems. Zero-valent iron nanoparticles have been applied in soil remediation in the USA,[9] and sensors that rely on nanotechnology are being developed to monitor ecological change.[10] Nanocoatings to prevent soiling of windows and other surfaces reduce the need for detergents and hence the potential environmental damage caused by detergent use.

3.18 In a broader context it has been pointed out to us that some nanotechnologies will contribute to reduced energy use, waste minimisation and improved recycling capability, all beneficial to the environment. For example, the use of cerium oxide as a fuel additive reduces 'soot' formation and has been reported to improve fuel efficiency. However, evidence supplied to us by the Woodrow Wilson Center suggested that the manufacture of some types of other nanomaterials is energy intensive and is itself highly polluting. We have been told that in one process used for manufacturing fullerenes, only 10% of material was usable and the rest was sent as waste to landfill.[11]

3.19 Other examples of how the introduction of nanomaterials may benefit the environment can be found in improved monitoring devices that are less expensive and more sensitive than current devices. New protein-based nanotech sensors make possible the detection of mercury at very low concentrations (one part in 10^{15} or one quadrillionth),[12] while nanoparticulate europium oxide can be used to measure the pesticide atrazine in contaminated water.[13] Nanotechnology has also improved the monitoring of atmospheric pollutants by utilising thin layers of nanocrystalline metal oxides as crucial components of solid state gas sensors. Measuring small changes in electrical conductivity allows detection and quantification of methane, ozone and nitrogen dioxide.[14]

31

Novel toxicological threats

3.20 In Chapter 2 we noted that many kinds of manufactured nanomaterials are considered to be functionally novel because their physical, chemical and biological characteristics differ from those of the same substance in bulk form. It follows that if the properties are new and unexpected then there is also potential for new and unexpected toxicological effects to emerge. To repeat the key message in Chapter 1, it is the functionality of novel materials in what they do and how they behave that matters, not their size *per se*.

3.21 To assess the relative safety of nanomaterials it is no longer possible to rely on the health and safety information developed for their bulk counterparts. The limited toxicological information which is available for specific nanomaterials is rarely put in context by comparison with the toxicity of the same material in bulk form. Researchers and manufacturers, who have harnessed a specific property of a nanomaterial for a particular purpose, have sometimes been surprised to see other functional properties emerge which they had not expected, or could not explain, even within the controlled environments under which these materials were developed and tested.[15] Newly-emerging properties are even more problematic when attempting to assess nanomaterial behaviour in more complex real world situations. As a consequence, there is a compelling case for potential environmental and health risks to be identified at every stage of the life cycle of any nanomaterial to be used in the development of a specific product (see Chapter 2).

3.22 Our Twenty-fourth Report, *Chemicals in Products: Safeguarding the Environment and Human Health*, pointed out that the historical record is replete with unexpected toxicological impacts arising following the use of anthropogenic chemicals. We have learnt a great deal from these early episodes (see Chapter 1). If chemicals are known to be persistent and also bioaccumulate, then there are controls in place to carefully manage them by restricting their release into the environment. However, we may still be caught unawares, as witnessed with the emergence of a large number of different endocrine disrupting chemicals during the 1980s and 1990s. It was not foreseen that low concentrations of chemicals used as antifouling agents (tributyltin), surfactants (nonyl phenol), flame retardants (polybrominated diphenylethers) and plasticisers (phthalates) would bind to hormone receptors or disrupt hormone metabolism in birds, reptiles, fish and invertebrates, and possibly influence sperm counts and the development of testicular malignancy in humans.[16]

3.23 These examples refer to chemicals whose reactivity it was felt was reasonably well understood. This is not the case with many manufactured nanomaterials, for which almost nothing is known of their potential environmental effects or their likelihood of causing unintended harm. With earlier pollutants it was also possible to detect and quantify their presence in ecosystems and organisms. Measurement techniques for the nanoforms of materials in environmental samples do exist, but are cumbersome, time consuming and are not widely available. As already noted (2.26), nanomaterials are now reportedly used in over 600 products[17] and yet there is little or no knowledge of their life cycles or ultimate fate in the environment.

Nanotoxicology

3.24 All living organisms are exposed to toxicological threats from the environment. These have been combated through the evolution of a battery of defences, including barriers to uptake, entrapment and removal in secretions, the generation of factors that neutralise or break down the substance to aid elimination (e.g. antibodies and detoxification enzymes) and cells capable of ingestion, digestion and sequestration. These defence systems are effective to varying degrees and are versatile, but can be overwhelmed by highly toxic chemicals, by high-level exposure or by low-level chronic exposure (as in the case of air pollution) and by chemicals with novel properties. As with bulk forms of chemicals, organisms have been exposed to a wide variety of naturally-occurring nanoparticulates during evolutionary history, in the form of volcanic emissions, combustion products, dusts, viruses, and pollen, fungal or bacterial fragments. Nanomaterials arising naturally appear to be dealt with effectively by most organisms. Manufactured nanomaterials with enhanced specific chemical reactivity might exceed the ability of defence systems to cope. Indeed, this quality is capitalised on in the design of nanomedicines to ensure that drugs are delivered to particular cell types, and even to particular sites within cells, without initiating immune defence responses (box 2A).[18] It is these very properties that are of concern when nanomaterials are taken up unintentionally by non-target organisms.

3.25   A few manufactured nanomaterials have been used for long periods without apparent harmful effects on humans, the environment and other living organisms (e.g. titanium dioxide or zinc oxide as sunscreens), but for new nanomaterials now being produced, there is very limited or no toxicological information. These include carbon nanoparticles (fullerenes), nanometals (nanosilver, nanogold, etc.), carbon nanotubes, nanofibres constructed from other elements (magnesium, aluminium, manganese, etc.) and nanoparticles of one kind doped with other elements. Managing nanomaterials in the face of this ignorance poses an enormous challenge.

ASSESSING THE POTENTIAL ADVERSE ENVIRONMENTAL AND HUMAN HEALTH EFFECTS OF NANOMATERIALS

3.26   Ecotoxicology is the study of the fate and effects of anthropogenic chemicals (and radiations) on ecosystems and their component organisms.[19] In preparing this report, with one very recent exception,[20] we have not become aware of any ecotoxicological research addressing the effects of manufactured nanomaterials on ecosystem structure or processes, or on populations and communities of organisms *in situ* other than micro-organisms.

3.27   Studies of the ecotoxicological fate and effects of manufactured nanoparticles and nanotubes are in their infancy, with many researchers still discussing the suitability or not of conventional toxicological test procedures and risk assessment methodologies embraced within the EU system of regulation on chemicals and their safe use, **R**egistration, **E**valuation, **A**uthorisation and Restriction of **Ch**emical substances (REACH) (the requirements of REACH are discussed in more detail later, 4.20-4.34). Current ecotoxicological studies have focused principally on acute toxicity in aquatic species. With the exception of air pollution by particulates which have not been deliberately manufactured, little work has been undertaken to determine the effects of nanomaterials in water, soils, sediments or the atmosphere.

3.28   It is of some concern that of the relatively few studies undertaken to assess the ecotoxicology of manufactured nanomaterials, many have been inconclusive. This is evident in difficulties over whether or not an observed adverse effect was caused by the nanoparticles themselves, by a coating or other acquired properties, or was attributable to the transport medium. An example of this difficulty is research that investigated the toxicity of C_{60} fullerenes and reported oxidative injury in brains of fish,[21] but failed to adequately account for the effects of the tetrahydrofuran vehicle used to generate the aqueous aggregates. Subsequent work demonstrated that C_{60}

aggregates retain the solvent,[22] with further tests demonstrating that toxicity may be associated with tetrahydrofuran decomposition products rather than C_{60} itself,[23] although, physico-chemical characterisation was limited.[24]

3.29   The information provided about nanomaterials in terms of particle size distribution or other important physico-chemical properties is frequently inadequate, reducing the possibility of repeating the work in other laboratories or comparing the results of one study with another. This is important as these characteristics can change, for instance as molecules agglomerate during storage. We are of the view that this lack of attention to detail in many (but not all) current studies greatly undermines confidence in the reliability of the conclusions that can be drawn from the work.

3.30   In contrast to ecotoxicologists, toxicologists studying human health impacts can to some extent draw on the experience gained from traditional toxicological and occupational exposure investigations into the adverse effects of exposure to dusts such as asbestos, silica and to atmospheric particles derived from combustion processes (appendices G and H). However, discounting these inhalation studies, nanotoxicology is still very much in its infancy.

3.31   From the extensive evidence we have received, as well as from our visits in the UK and overseas, we are not aware of any evidence of severe adverse effects of manufactured nanomaterials on ecosystems or their component organisms *in situ*. This is consistent with the assessment in a report for the Department for Environment, Food and Rural Affairs (Defra) conducted by the Central Science Laboratory (CSL) York and associates in 2007,[25] which concluded that exposure estimates based on current use and production patterns in the UK were many orders of magnitude lower than the concentrations likely to cause acute effects in invertebrates, fish or algae, or sublethal effects on fish, invertebrates or bacteria. However, the report noted "Whilst this study has identified the potential environmental exposure arising from a range of key ENP (Engineered Nanoparticle[i]) types, the assessment has been limited by the availability of data and knowledge. Work in the future should therefore focus on: 1) establishing a detailed knowledge of the content and use of products containing ENPs in the UK; 2) developing an understanding of the factors and processes affecting the fate and transport of ENPs in the environment; 3) the development and evaluation of more complex exposure assessment models; and 4) the development of a better understanding of the ecotoxicity of ENPs under environmentally-relevant exposure situations". We generally concur: this remains an area of great uncertainty. We also note that this study relates to a narrow range of effects. We do need to address the question "Would we know if nanomaterials were causing damage?" in a wider context. With our extremely limited understanding regarding exposure levels and patterns, as well as our ignorance of the toxicology of nanomaterials, we cannot be confident of knowing whether effects are occurring or will in future occur in the wider environment.

BIOLOGICAL DAMAGE FOLLOWING EXPOSURE TO NANOMATERIALS

3.32   Currently, the scientific literature on nanotoxicology consists of around 800 publications, the vast majority of which concern the cytotoxic effects of nanomaterials in cell culture systems – mostly mammalian and involving transformed or tumour-derived cell lines. However, these *in vitro* studies are a long way from giving us an understanding of the toxicity of manufactured

---

i   The term 'Engineered Nanoparticle' as used in the CSL report is broadly equivalent to the term 'manufactured nanoparticle' used in this report.

nanomaterials in nature. Almost all the evidence presented to us related to comparative toxicology and toxicity testing in the laboratory, rather than systematic investigations of how nanomaterials might affect growth, reproduction, viability of offspring of various species, population dynamics or community structure in the real world. On a larger scale, we came across no conclusive investigations of how nanomaterials might affect ecosystem structure or processes. However, while still at an early stage, we are encouraged by a rapid growth of interest in this field.[26]

3.33   Although nanoparticles and nanotubes have diverse properties and fall within a relatively wide size range (1-100 nm in at least one dimension), some common toxic mechanisms may be associated with different kinds of nanomaterials. Toxicity due to the generation of reactive oxygen species is frequently attributed to nanomaterials, giving rise to effects on cell membranes, cytoplasm, nuclei and mitochondrial function.[27] However, it is not always clear whether such damage in a complex organism is due to the direct effect of the nanomaterial or to an indirect effect attributable to an inflammatory response involving the influx of secondary cells, induced by the nanomaterial entering tissues.[28] Although popular as a unifying mechanism, it remains possible that oxidative stress has been overplayed and represents a one-dimensional approach to a very complex and multi-factored problem.

3.34   Nanotoxicity has been related to the capacity of nanoparticles and nanotubes to act as vectors for the transport of other toxic chemicals to sensitive tissues (the Trojan Horse effect). In a study with carp, cadmium accumulation was increased 2.5-fold when titanium dioxide nanoparticles were added concurrently with cadmium salts.[29] This Trojan Horse mechanism was also seen when aggregates of fullerenes and a representative range of organic contaminants were investigated.[30] The toxicity of phenanthrene to algae and to water fleas (*Daphnia magna*) was increased following sorption to C_{60} aggregates, attributed to the delivery of the phenanthrene directly to cell membranes. However, the toxicity of pentachlorophenol decreased when associated with C_{60} in the form of aggregates.

3.35   Knowledge derived from the behaviour of particles and from colloidal chemistry in environmental media can aid the prediction of nanoparticle behaviour,[31] and the ways in which manufactured nanomaterials agglomerate, aggregate, disperse, adhere to surfaces and other particles, and interact with natural organic matter (see Chapter 2) are all of relevance.[32]

3.36   The factors used to predict environmental risk, in combination with production volume, include persistence, bioaccumulation and toxicity. The persistence of different classes of manufactured nanomaterials is likely to vary widely. Evidence given to the Commission in the US asserted that carbon nanomaterials were likely to be highly persistent. Indeed, carbon nanotubes are some of the least biodegradable man-made materials known,[33] are insoluble in water and are lipophilic (i.e. have a preference for entering fatty cell membranes).[34] These characteristics are normally associated with a tendency to bioaccumulate, and might indicate carbon nanotubes are likely to bioaccumulate in food chains and be highly persistent.[35] However, of some concern is that almost nothing is known about the stability of other kinds of manufactured nanomaterials.

3.37   As reported above (3.31), current estimates of exposure levels are low but, with exponential growth in this manufacturing sector (see Chapter 2), greater quantities are likely to be released in the future. There exists a broad consensus that a great deal more needs to be known about the fate, persistence and transformation of manufactured nanomaterials in the environment and the mechanisms by which toxic effects are produced at particular exposure levels.

3.38 Some of the early findings from laboratory studies of the effects of nanomaterials on micro-organisms, plants, invertebrates, fish and other wildlife are summarised below. More details can be found in recent reviews.[36, 37]

3.39 Possible mechanisms of nanoparticle uptake into bacteria are non-specific diffusion, non-specific membrane damage and specific uptake.[38] The largest globular proteins shown to pass through the cell wall of *Bacillus subtilis* had a radius of 2 nm; it is unlikely that particles larger than this would enter bacteria in significant quantities by non-specific diffusion. Using electron microscopy quantum dots of < 5 nm have been shown to enter *Bacillus subtilis*.[39] Entry through damaged bacterial membranes has been demonstrated for highly reactive manufactured nanomaterials such as halogenated nanoparticles. Endocytosis as an active process of particle uptake also cannot be ruled out as a possible mechanism (as described in mammalian cells, appendix I). Once inside the organism, the bactericidal activity of silver particles and titanium dioxide has also been extensively recorded, including detailed studies of the mechanisms of antibacterial activity and the use of photosensitisation to augment the formation of reactive oxygen species.

3.40 There is little information available regarding the transformation of nanomaterials by microbes.[40] Reduction–oxidation reactions are often mediated by micro-organisms, either directly through enzymatic activity or indirectly through the formation of biogenic oxidants or reductants.[41] Biological modifications, as well as degradation of the surface properties of nanoparticles, may result in modification of structure and, as discussed already, the release of entrained constituents such as metals and solvents.

3.41 It is difficult to extrapolate these studies to complex microbial ecosystems in the absence of information about the environmental fate and behaviour of nanomaterials. One study found that fullerenes in various forms had no effect on soil respiration,[42] and a follow-on study confirmed the lack of effects of C_{60} when used as a substrate for anaerobic sludge digesters.[43]

3.42 Mechanisms allowing manufactured nanomaterials to pass through cell walls of algae and fungi are not well understood.[44] However, once inside cells, nanomaterials behave similarly to how they behave in higher organisms; effects include physical restraints (clogging effects), solubilisation of toxic compounds and production of reactive oxygen species.

3.43 Mycorrhizal fungi are also adept at taking up metals and can be important for sourcing nutrients, for example zinc, for the host plant. Whether they can take up nanoforms, such as zinc oxide, is unknown. Fungi themselves are important for bioremediation and a recent report has shown their utility in recovering depleted uranium from military activities.[45] We have every reason to believe that fungi would be able to take up nanometals or their compounds, but to our knowledge this has not yet been tested.

3.44 One study has described the formation of nanocrystals of cadmium on phytoplankton.[46] The toxicity of silver nanoparticles to the marine diatom *Thalassiosira weissflogii* has also been investigated under different nutrient conditions using growth rates and photosynthesis as toxicity endpoints. A near linear relationship between toxicity and the release of silver ions from the particles has been reported.[47] Plant tissues might serve as scaffolds for aggregation of metallic nanoparticles *in situ*.[48] Thus, carbon nanotubes can be taken up by microbial communities and by root systems to accumulate in plant tissues.[49] However, the behaviour of manufactured nanomaterials in higher plants is largely unknown, despite their ecological importance and the potential mechanism which they provide as a vehicle for introducing nanomaterials into the food chain. As noted in our short report *Biomass as a Renewable Energy Source*, some plants, such as willows, can take up heavy metals (e.g. cadmium) from contaminated soils.[50] If they can take up bulk cadmium, it is highly plausible that they can also take up nanoforms of the element or its compounds.

3.45 Studies involving exposure of invertebrates to manufactured nanomaterials have largely followed current OECD guidelines for chemical testing, starting with basic testing for acute toxicity in aqueous media. These tests often only include one invertebrate species, the water flea *Daphnia magna*. In view of both inter-specific and intra-specific variability in response to exposure to nanomaterials (and also bulk chemical forms), it is difficult to envisage how conventional single species toxicity tests can provide sufficient information on which to base effective measures for the protection of the environment. Single species toxicity tests have been performed with nanomaterials using a variety of test organisms ranging from bacteria and algae to benthic invertebrates and fish.[51, 52] The data derived confirm that different species, even within a taxonomic genus, can show very different sensitivity to different nanomaterials.

FIGURE 3-II

Nanoparticulate uptake by *Daphnia magna*[53]

The picture shows uptake of polystyrene beads in D. magna. *The beads are labelled with a fluorescent dye and observed by confocal microscopy. It can be seen that the fluorescence accumulates in small oil storage droplets, having already passed the epithelial gut barrier.*

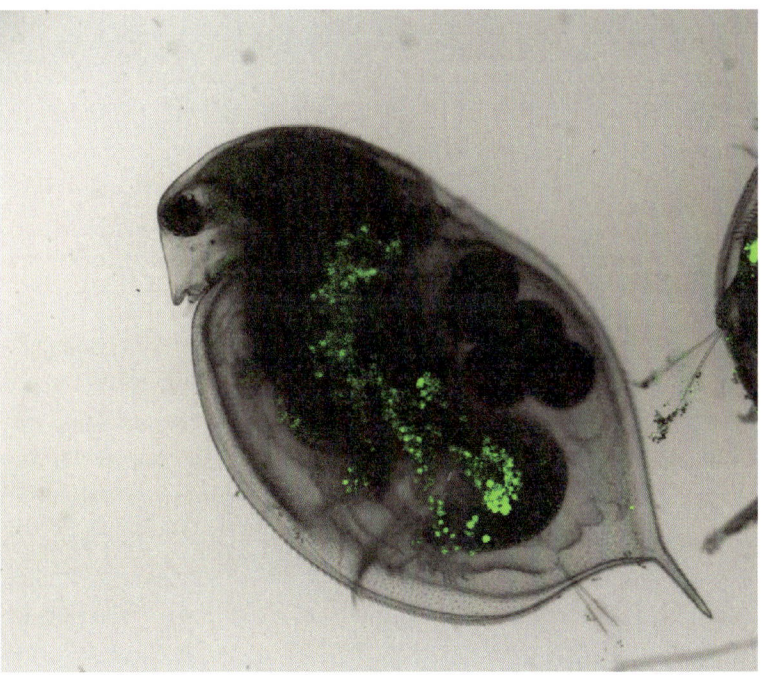

Reproduced by kind permission of Professor Vicki Stone, Napier University, Edinburgh.

3.46 Testing nanomaterials in aqueous media often results in aggregation, which influences both uptake and toxicity and also confounds interpretation of the toxicity test results. Despite this, there is evidence for rapid uptake of carbon black, titanium dioxide and nano-sized polystyrene by *Daphnia magna* (figure 3-II) under standardised laboratory conditions.[54] Qualitative uptake of other types of nanomaterials, including carbon nanotubes, has also been demonstrated.[55] There is some evidence that *D. magna* can also modify the solubility of carbon nanotubes during aqueous exposures.[56] In general, oxide nanoparticles, including those of aluminium, silicon and titanium, exhibit low toxicity in acute and chronic exposure studies with *D. magna*.[57] Whether this is the case for representatives of other invertebrate phyla is not known.

3.47   In aquatic animals, nanomaterial uptake across gills and other epithelial body surfaces occurs.[58] Thus, it is somewhat surprising that greater attention has not been given to the effects of nanomaterials on detritivors and filter-feeders which ingest large amounts of particulate matter and consequently are most likely to encounter nanomaterials and concentrate them from water.

3.48   Early life stages appear to be particularly sensitive to toxicants, including manufactured nanomaterials. Zebrafish embryos exposed to single-walled carbon nanotubes revealed reduced hatching success.[59] In a study with Japanese Medaka fish, fluorescent nanomaterials accumulated in various organs and were able to pass through the blood–brain barrier.[60] An *in vivo* study of quantum dot uptake by embryos of the amphibian *Xenopus* showed that internalised quantum dots could be transferred to daughter cells upon cell division.[61] This may have important implications for transgenerational effects of nanomaterials, but these have yet to be considered.

3.49   There has been some study of the interactions between dissolved and component metals and manufactured nanomaterials in fish. The accumulation of cadmium in carp has been investigated in the presence and absence of titanium dioxide nanoparticles or sediment particles.[62] Sediment particles alone had no effect on the uptake of cadmium, but when titanium dioxide nanoparticles were present bioaccumulation of the metal was observed. This indicated that titanium dioxide nanoparticles had a higher adsorption capacity for cadmium than natural sediments and that the metal and nanoparticles could accumulate in the viscera and gills of fish.[63] Exposure studies that have been performed with fish have been hindered because high concentrations of nanomaterials in aqueous media aggregate, resulting in reduced uptake. Under these conditions it is difficult to determine lethal concentrations. In addition, there are profound problems inherent in the use of liquids for dispersing particles (3.28).[64]

3.50   Disappointingly, there are few studies of how manufactured nanomaterials are dealt with by terrestrial wildlife species, other than studies on laboratory rodents. As we discuss in the section on mammalian toxicology below (3.57), ingestion and inhalation are likely to be the major routes of uptake into terrestrial organisms.[65, 66] As with other unexpected toxicity problems in the past, careful study of mammalian wildlife species is likely to provide greater understanding of potential toxic threats to humans.

3.51   The issue of bioaccumulation and entry of nanoparticles and tubes into the food web has yet to be seriously addressed. A preliminary study of single-walled carbon nanotubes ingested by the nematode *Caenorhabditis elegans* showed movement of the nanotubes through the digestive tract. They did not appear to be absorbed by the animals,[67] but their presence in the gut suggests the possibility of entry into the food web if the nematodes were subsequently ingested by other animals.[68]

3.52   Chronic, full life cycle testing in aqueous and sediment compartments is a research imperative if the ecotoxicity of manufactured nanomaterials and associated contaminants is to be better understood. Evidence has been presented to us that many nanomaterials, including fullerenes and silver nanoparticles, are likely to persist in the environment and therefore remain bioavailable. Chronic effects on growth, reproduction and viability of offspring are of particular concern and might ultimately affect inter-specific relations and the functioning of multi-species systems. It may be more illuminating to carry out basic tests with several species of organisms, perhaps from different taxa or representing different feeding types, reproductive strategies, etc., in order to gain at least some indication of the natural variability of response, rather than performing very precise testing with just one species. As is the case in toxicity testing of bulk chemicals, the ecological relevance of the toxicity data is extremely limited.

3.53   Biomarkers might be of value in assessing exposure to and effects of nanomaterials both in the laboratory and *in situ*.[69] For example, we were told that the metal-binding protein metallothionein responds to the presence of cadmium-containing nanomaterials, signalling their presence,[70] but little progress has been made in this field. We believe that more work of this nature is required.

3.54   Investigating gene expression profiles (sometimes called toxicogenomics) as a way of evaluating the sublethal effects of materials has been proposed.[71] In fibroblasts exposed to coated cadmium selenide and zinc sulphide quantum dots, about 50 genes in a microarray were differentially expressed.[72, 73] These microarray techniques offer a novel and comprehensive way of identifying potential biomarkers and suites of biomarker responses to nanomaterials in a wide range of animals, plants and micro-organisms. However, as noted in our report *Chemicals in Products*,[74] this kind of approach is not well developed even for conventional chemicals and well-known organisms, but it holds considerable promise as a new toxicological tool.

3.55   The absorption, distribution, metabolism, excretion and toxicity of carbon nanotubes in organisms depend on the inherent physical and chemical characteristics, such as charge transfer, functionalisation, coating length and agglomeration or aggregation state, that are influenced by external environmental conditions during production, use and disposal (see Chapter 2). Characterised exposure scenarios could therefore be useful when conducting toxicological studies.[75] The likelihood of organisms being exposed to nanomaterials may vary greatly depending on where they live, their feeding and reproductive strategies and their behaviour. Some types of nanomaterials are also more likely to enter the environment in greater amounts than others. From the evidence that we have received, the greatest concerns at present relate to fullerenes, single-walled and multi-walled carbon nanotubes and nanosilver.

3.56   All this being said, and despite plausible mechanisms and pathways by which manufactured nanomaterials might harm organisms and ecosystems, recent analyses suggest that even when using "highly conservative estimates of exposure, concentrations in the environment are likely to be considerably lower than concentrations required to produce toxicological effects".[76] But with exponential growth in the types of nanomaterials and their applications (see Chapter 2) it is unclear how long this state of affairs will continue.

THREATS POSED BY NANOMATERIALS TO HUMANS

Exposure routes and uptake of nanoparticles in humans

3.57   From the evidence we have examined it appears that a great deal more is known about exposure to and toxic effects of nanomaterials in rodents (as 'model organisms' for human toxicology assessments) and of particles and dusts in humans than for any other species (appendices G and H). In the following sections this toxicological information is examined in greater detail. Experience gained from the study of occupational dust diseases and the epidemiology for atmospheric particulate pollution and human and animal health suggests that inhalation is the most likely route of unintended exposure to manufactured nanomaterials. Air pollutants as well as manufactured nanoparticles released into the atmosphere may stay suspended for considerable periods without agglomerating. At high concentrations, agglomeration can be very rapid but is dependent upon the charge characteristics of the particles. Other routes of entry such as the skin and gastrointestinal tract have more robust exclusion mechanisms; skin presents a physical barrier and mechanisms in the gut are designed to absorb, transport and process small particles, for example, fat is absorbed across the intestine as small particles (chylomicrons) that form in the presence of bile salts.

Inhalation exposure and particle uptake

3.58   The human lung is made up of multiple fractal divisions of conducting airways that deliver inhaled air into the small air sacs (alveoli) at the lung periphery. Here, oxygen is efficiently exchanged for carbon dioxide across a total surface area the size of a tennis court (or 30 times the surface area of the skin).

3.59   Inhaled particles 10-100 μm in aerodynamic diameter are mostly trapped in the nose through impaction on the mucous membrane. If this fails (e.g. with mouth breathing) or with particles $\leq 10 \mu$m, particles pass into the lung where turbulence at airway bifurcations leads to impaction onto the airway lining aided by surface secretions containing mucus. The 'mucociliary escalator' then transports the particles entrapped on the mucus layer to the oropharynx where they are swallowed. These mechanisms reduce the majority of large and intermediate-sized particles from reaching the delicate alveoli at the periphery of the lung,[77] although a small proportion of these particles will do so. As the diameter of the particles becomes smaller, especially those < 100 nm (i.e. nanoparticles), an increasing proportion stay suspended, aided by repellent electrostatic forces, and are deposited in the alveoli (figure 3-III).[78] Smaller particles (0.1-1 nm) fail to deposit and may be exhaled (e.g. as occurs with a proportion of particles in exhaled tobacco smoke).

3.60   Once in the alveoli, the primary mode of particle removal is through uptake (phagocytosis) by scavenger cells (mostly macrophages) (figures 3-IV and 3-V), but when these cells are overwhelmed, particles are taken up by white blood cells (neutrophils) that migrate into the air spaces from the blood stream.[79] Together, these scavenger cells either break down the particles using an array of intracellular enzymes or, if not biodegradable, carry them up to the mucociliary escalator for subsequent elimination by swallowing.

3.61   Under normal circumstances these mechanisms are sufficient to cope with particles inhaled from an ambient environment by the lungs. However, macrophage particle uptake is a saturable process and at high particle loads it becomes inefficient. Heavy particle loads and changes to the properties of particles lead to the secretion of tissue-damaging mediators and enzymes with the capacity to cause inflammation and fibrosis.[80] Inefficient particle removal by scavenger cells, as occurs with high concentrations of nanoparticles, facilitates their passage through the delicate walls of the air sacs into the surrounding lung tissue and small blood vessels that surround them. In these situations they are treated as invading foreign bodies (like micro-organisms) exciting an inflammatory and/or fibrotic response (figure 3-IV).

FIGURE 3-III
Fractional deposition of inhaled particles[81]

Model of fractional deposition of inhaled particles ranging from 0.6 nm to 20 μm in the nasopharyngeal/laryngeal (NPL), tracheobronchial (TB) and the alveolar (A) regions of the human respiratory tract during nasal breathing. Within the ultrafine particle size range (i.e. less than 100 nm) there are significant differences in each of the three regions with regard to their deposition probabilities.

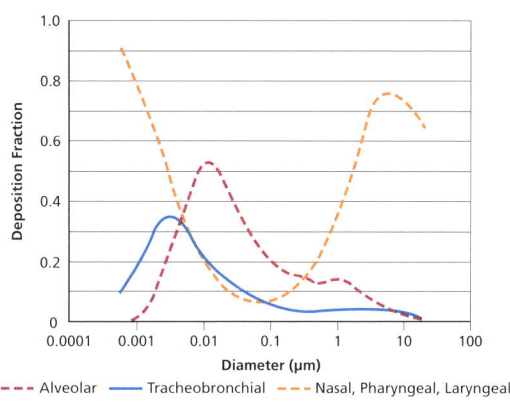

FIGURE 3-IV
Nanoparticle uptake by lung macrophage[82]

Schematic representation of nanoparticle uptake by lung macrophages, the release of inflammatory mediators and the passage of particles through the surface epithelium.

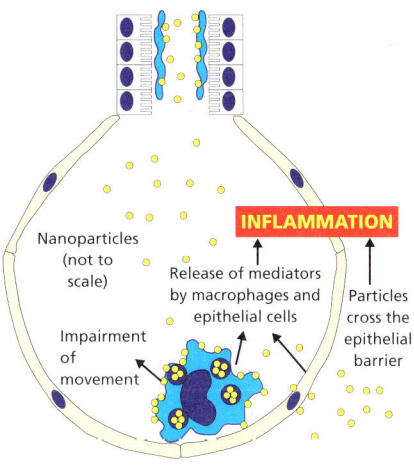

41

FIGURE 3-V
Lung macrophage in lung tissue of infant[83]

Transmission electron micrograph of a lung macrophage in lung tissue from an infant exposed to ambient air pollutant particles showing their uptake into small intracellular vesicles.

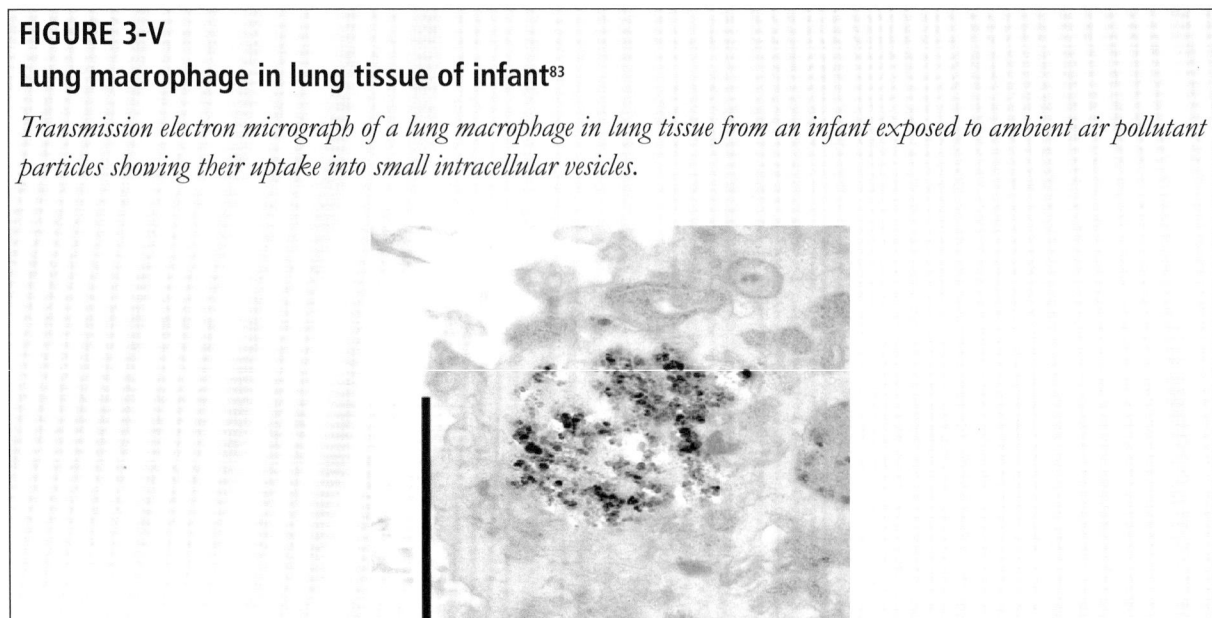

3.62 An increased understanding of the processes involved in the passage of nanoparticles from the surface of the air sacs into lung tissue has come from studying the potential of the inhaled route to deliver certain drugs systemically (such as dry powder formulations of insulin[84]) and the fate of inhaled drugs such as bronchodilators or corticosteroids used in the treatment of common lung disease. Most inhaled drugs gain access to the circulation by passing between adjacent epithelial cells (paracellular transport), although some are designed to pass into the cells (transcellular transport) (figure 3-VI).

FIGURE 3-VI
Movement of particles between epithelial cells

Passage of particles between adjacent airway epithelial cells by negotiating tight junctions (A) or disrupting them (B) (paracellular transport) or by direct uptake into epithelial cells (C) (transcellular transport).

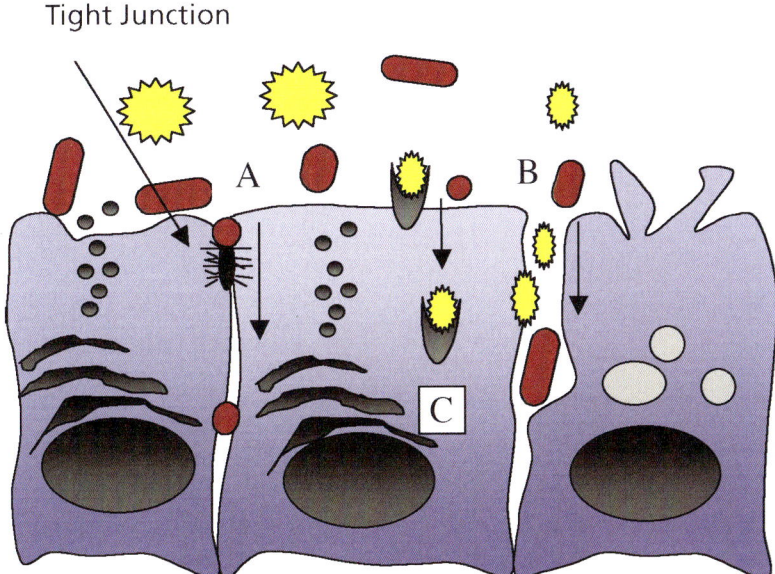

Image kindly provided by Professor Stephen Holgate

3.63   There is good evidence in rodent models that 0.3-0.5% of instilled nanoparticles can pass through the lungs and into the circulation, ultimately being trapped by scavenger cells in the liver, bone marrow and spleen.[85] The ability of nanoparticles to pass through the epithelial lung barrier is highly dependent upon their specific composition, shape and surface charge.[86] Inhalation studies of nanoparticles for humans are limited to only three studies using Technegas®, comprising 5-100 nm technetium (Tc)[99m]-labelled carbon particles, which yielded conflicting results because of the dissociation of the radiolabel from the particles.[87] Moreover, inhalation of carbon nanoparticles by normal and asthmatic subjects produced only transient retention by circulating white blood cells in the lung vasculature, but no evidence of systemic effects.[88]

3.64   In rodents, and also possibly in humans, nanoparticles can pass from the nasal mucosa to the brain via the olfactory bulb, where they are capable of exciting an inflammatory response[89] (figure 3-VII). Nanoparticles have also been shown to pass across human nasal epithelium *in vitro*,[90] but whether this pathway bypassing the blood–brain barrier occurs in humans is unknown, although polio virus can gain access to the central nervous system via this route.

Gastrointestinal uptake

3.65   With regard to this route of particle uptake, the gastrointestinal epithelium utilises both paracellular and transcellular pathways. In rats fed for 10 days with 50 nm I[125]-labelled polystyrene microspheres, over one-third of the particles were absorbed by and transported via the lymphatics to the liver and spleen.[91] If the mucosa is breached by a disease process such as inflammatory bowel disease, uptake of particles will increase. However, the gastrointestinal fluids are high in ionic strength that will encourage particle aggregation. On reviewing the available published literature it seems that most effort has been focused on utilising nanotechnology for delivering biopharmaceuticals and diagnostic agents rather than investigating the possible toxicity of ingested nanomaterials (box 2A).

FIGURE 3-VII
Human nasal passage system

Based on studies in rodents, passage of nanoparticles (NP☀️) from the nose into the brain via the cribriform plate that separates the nasal cavity from the brain and supports the olfactory nerves and receptors (olfactory bulb, green) responsible for sensing smell and taste.

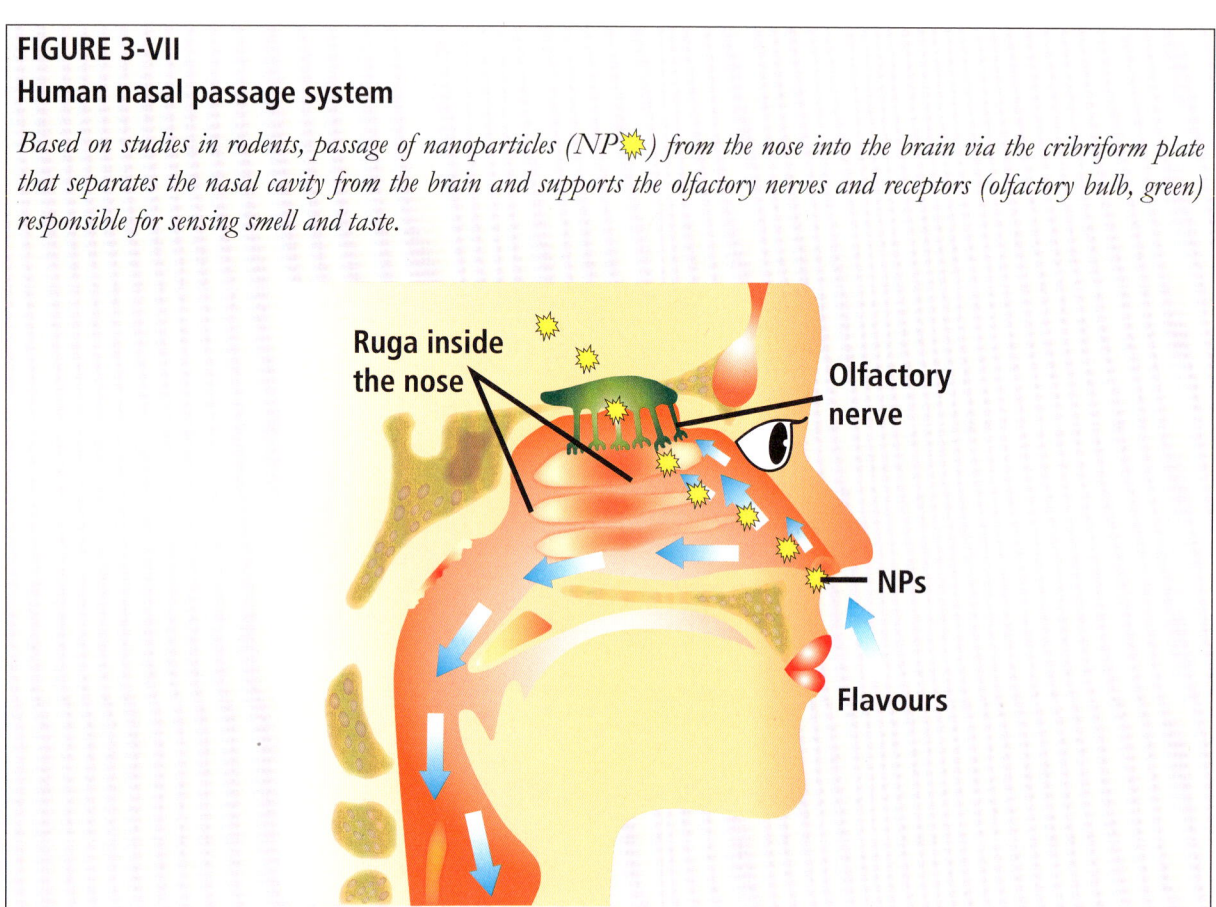

Uptake through the skin

3.66 The skin provides a highly effective barrier to nanoscale particles which are increasingly being used in cosmetics and sunscreens (e.g. titanium dioxide and zinc oxide) and which are generally considered safe.[92] In mice, however, polymer nanoparticles can penetrate the skin to cause, in the case of beryllium, systemic allergic sensitisation.[93] Cadmium selenide quantum dots (QDs) of 4.6 nm diameter penetrate the deeper parts of the skin with the particle surface coating being a primary determinant of toxicity.[94]

3.67 The recent discovery of genetic variants in proteins that are involved in maintaining skin barrier function, such as those encoded in a cluster on chromosome 1q (epidermal differentiation complex), raises the possibility that particular individuals could be especially vulnerable to increased particle uptake and, as a consequence, absorption and toxicity. Loss of function genetic mutations in the filaggrin gene occur in up to 9% of the population and cause both dry skin and a strong predisposition to eczema (atopic dermatitis).[95] In the respiratory epithelium, epithelial integrity is also broken in asthma through defective formation of tight junctions and epithelial damage which may account for the increased systemic uptake of particles in this disease.[96]

Factors determining the mammalian cellular toxicity of nanoparticles

3.68   In trying to assess the potential of manufactured nanoparticles to cause adverse effects in humans (and indeed other organisms), it is important to understand the relationships between their physical and chemical structures and their biological effects. Two criteria have been proposed to identify nanomaterials which may present a unique potential risk to human health:[97]

- the material must be able to interact with the body in such a way that its nanostructure is biologically available; and

- the material should elicit a biological response associated with its nanostructure different from that associated with non-nanoscale material of the same composition.

3.69   New materials present new challenges to evaluation of risk. The risk assessment of organic chemicals also presented a challenge to which the development of quantitative structure–activity relationships (QSARs) helped bring order. But the critical question is whether QSARs can be developed for nanomaterials. In Chapter 2 we introduced the important concept that the biological effects of nanoparticles once inside the cell are critically dependent on their size, shape and composition. There is, therefore, a critical need to link particle properties to hazard. However, the multiple dimensions of nanomaterials (chemical, physical, biological and commercial) expose a new kind of complexity compared to organic chemicals. While a long list of factors are known to influence cellular responses, some are of greater importance than others.

3.70   *Mass and volume versus surface area* It has been demonstrated that, for the same mass, instillation of titanium dioxide as 25 nm particles into the lungs of rats produced a much more severe inflammatory response than achieved with 250 nm particles.[98] Using a variety of different nanomaterials, surface area has been shown to be the most important generic factor driving epithelial-induced inflammatory as well as cardiovascular responses.[99] However, with other nanomaterials, mass and size may be at least as important as other factors for two reasons: the ability to penetrate barriers, e.g. cell walls and membranes; and the possibility of direct molecular interactions with biological molecules (e.g. changing their configurations), something not possible for larger particles.

3.71   *Surface properties* As might be predicted from the relative pulmonary toxicity of different minerals (appendix G), the surface characteristics of nanoparticles exert a major influence on the nature and magnitude of ensuing cellular responses. One example is the greater pulmonary toxicity of nanoparticles comprised of silica compared to those of titanium dioxide or barium sulphate.[100] The pulmonary toxicity of alpha-quartz particles correlates more closely with surface activity than either particle size or surface area.

3.72   *Particle shape* Another critical factor determining particle toxicity is shape. Single-walled carbon nanotubes (SWCNTs) have their own distinct morphology but also assemble into complex larger structures. In both mice and rats, SWCNTs, irrespective of how they are made, are more toxic to the lung than is quartz, producing both inflammatory and fibrotic responses.[101] Moreover, improved dispersion of aspirated SWCNTs results in increased uptake into mouse lung tissue, granuloma formation and increased fibrosis.[102]

3.73 Of great concern is the possibility that carbon nanotubes may predispose to mesothelioma, as is the case with asbestos fibres. Rodent models currently used in toxicity testing do not appear to be sufficiently predictive to reject this possibility.[103] Indeed, a recent report[104] has shown that exposing the peritoneal cavity of mice, as a surrogate for the mesothelial lining of the chest cavity, to long multi-walled carbon nanotubes resulted in asbestos-like behaviour. This included inflammation and the formation of collections of macrophages and immune cells (granulomas) considered to be a precursor of asbestos-related mesothelioma in humans.[105] As is the case with asbestos fibres, the extent of peritoneal inflammation was proportional to fibre length. These findings are clearly a concern and call for urgent further research. Currently there are no suitably 'predictive' tests that could be used to forecast the long-term toxic behaviour of nanomaterials.

3.74 *Particle composition* While particle mass may not be such an important factor in nanotoxicology, chemical composition is important. Exposure of human alveolar epithelial cells *in vitro* to nanoparticulate metals, such as silver, aluminium, zinc and nickel, and to titanium dioxide resulted in entirely different rank orders of potency-induced morphological damage, programmed cell death, generation of reactive oxygen radicals and nucleic acid fragmentation.[106] For nanoparticulate titanium dioxide, it is known that its photocatalytic activity[107] and its pulmonary toxicity are both dependent upon its crystal structure (anatase or rutile).[108] However, what is not clear is whether these differences are the result of different surface properties, shape or mass structure. Similarly, systematic investigation of iron- or other metal-containing nanoparticles, such as those used in industrial fine chemical synthesis, has revealed that the presence of catalytic activity strongly enhances cell cytotoxicity.[109]

3.75 For gold and titanium dioxide nanoparticles, there are important differences in intracellular localisation and inflammatory mediator release as a function of particle composition.[110] The use of well-characterised nanoparticles of the same morphology, comparable size, shape and degree of agglomeration has allowed separation of physical and chemical effects. Under these conditions, cobalt and manganese particles provoked up to an 8-fold greater oxidative stress signal when compared to cells exposed to aqueous solutions of the same metals and were far more active in this regard than similarly sized particles of iron or titanium dioxide.[111]

3.76 The majority of *in vitro* and *in vivo* studies on nanomaterials are conducted with single elements or compounds and not with the combinations of elements or compounds that are frequently used in their commercial application. In the case of SWCNTs, toxicity is greatly enhanced by the presence of trace metals.[112]

Mechanisms of toxicity in mammalian cells

3.77 With the epithelium being such an important site of nanomaterial absorption and toxic effects, attempts have been made to use *in vitro* cell culture systems to study uptake, transport and toxicity. While knowledge is accumulating rapidly in this field, most experiments have used transformed and tumour cell lines, but it is not at all clear how these cell lines compare to primary cells or differentiated cells. Nor is it known how different nanomaterials interact with various organ-specific versus circulating macrophages and how these responses compare to immortalised cell lines.

3.78 The principal damaging effect of intracellular nanomaterials most likely occurs through activation of oxidant pathways in which highly reactive hydroxyl and superoxide radicals are generated.[113] Oxidant damage to cell membranes is considered to account for a high proportion of the inflammatory and fibrotic effects of nanomaterials, although it has to be said that this mechanistic aspect has dominated research in this field. Particle penetration of the cell nucleus to interfere with chromosomal functions by oxidant pathways has also been used to explain toxic effects on cell division, including malignant transformation,[114, 115] but with further research, additional mechanisms are likely to be found.

3.79 We have already referred (3.34) to another important route to toxicity, namely the ability of nanomaterials to carry toxins into cells in a hidden form and then release these, perhaps in a concentrated form, to damage cellular machinery – the so-called 'Trojan Horse' mechanism.[116] This possible route exists both for the wider environment and for human toxicology.

COMPARING IN VITRO WITH IN VIVO MAMMALIAN TEST SYSTEMS

3.80 While much new knowledge is being gained from studying the fate and effects of nanomaterials in epithelial and other cells, the overall response of a complex tissue, such as the lung, involves interactions between numerous cells and cell types and must take account of dynamics involved in particle clearance. This will include dissolution, 'quarantining' (isolating particles to minimise their biological effects) and concentrating effects so that interacting factors become of major importance in generating chronic tissue injury and repair responses, as well as predisposing to neoplasia. Such investigations depend on whole animal studies.

3.81 When considering a range of toxicity endpoints for five different types of particle (carbonyl iron, crystalline silica, precipitated amorphous silica, nano-sized zinc oxide and fine-sized zinc oxide) comparisons of *in vivo* and *in vitro* measurements have revealed little correlation.[117] While a battery of *in vitro* tests could be used as an initial toxicology screen, the complexity of responses to nanomaterials means that for a wide range of toxic endpoints, especially when looking for chronic effects, *in vitro* tests cannot be substituted for whole animal studies.

3.82 While we fully recognise that any increase in animal testing is unacceptable morally, ethically and politically, the choice of whether whole animal tests should be undertaken will need to be greatly influenced by the risk–benefit ratio and a judgement by society over whether such tests are warranted. It was this consideration that led to the abandonment of animal testing for cosmetics. However, the case of multi-walled carbon nanotubes raises concerns similar to those for asbestos, and this might create adequate justification for *in vivo* testing, provided of course that there is an adequate case for the use of these nanomaterials for society's benefit in the first place. Clearly, a starting point is to prioritise those new nanomaterials which should be subjected to more detailed toxicological appraisal, based on their relevant physical and chemical properties and potential exposure scenarios.

3.83 An important feature of *in vivo* test systems is the ability of a target tissue (such as the lung) to accumulate nanomaterials, such as SWCNTs, thereby greatly prolonging exposure time. The retention of specific particle types in different lung compartments might explain differences in lung pathology. Another recently discovered complexity is the capacity of SWCNTs to reduce the ability of animals to fight off lung infections, a complexity that would not be picked up by *in vitro* tests.[118] There are also unforeseen endocrine and generation effects. In mice, pregnancy enhances lung inflammatory responses to otherwise relatively innocuous inert particles such as

47

titanium dioxide, while exposure of pregnant female mice to either inert or environmental air pollution particles results in increased allergic susceptibility in the offspring.[119]

3.84 For particulate air pollution, animal models have been critical in identifying potential mechanisms for adverse cardiovascular health effects in humans (appendix H). The intrapulmonary instillation of carbon black nanoparticles in mice genetically deficient in the protective low density lipoprotein (LDL) receptor and fed a high-cholesterol diet resulted in accelerated development of cholesterol deposits in the blood vessels (atheroma).[120] In Watanabe rabbits[ii] that naturally develop systemic atherosclerosis, inhalation exposure to ambient air particles promoted accelerated cardiovascular disease progression by recruiting inflammatory macrophages to atherosclerotic plaques.[121]

3.85 While evidence is still sparse, there are clues to suggest toxic actions of manufactured nanomaterials on a range of organs including the brain, liver, kidney, reticuloendothelial system (spleen, bone marrow and lymphatics) and the reproductive organs. The significance of these effects will depend on the systemic bioavailability of the specific nanomaterials.

3.86 In summary, these mammalian studies and the ecotoxicological work presented above (3.20-3.56) demonstrate that some manufactured nanomaterials do interact with biological systems and whole organisms to disturb normal function and produce damage. It is important, therefore, that procedures are made available by which both the hazards and risks posed by manufactured nanomaterials can be accurately assessed. These assessments must answer questions about how much material is likely to cause an adverse effect, how organisms might be exposed in practice and how the level of such exposure is related to the dose likely to cause an adverse affect. We now consider these issues.

RISK ASSESSMENT PROCEDURES

Current testing methodologies

3.87 The development of manufactured nanomaterials has provided a number of challenges for the standard environmental toxicological test methods and tools. During our study a number of individuals and organisations informed us that the traditional approaches to toxicological testing were appropriate for nanomaterials and that unexpected biological effects had not been seen.[122] Others gave opposing evidence.[123] On the basis of the evidence received we have concluded that current test procedures for assessing the risks posed by manufactured nanomaterials to the environment, non-human organisms and human beings are inadequate.

3.88 This statement requires some elaboration. In principle, for certain classes of nanomaterials in widespread use (e.g. titanium dioxide), toxicological testing procedures to protect human health appear to be appropriate, but ecotoxicological work is conspicuous by its absence. Also, for many kinds of novel nanomaterials there are no agreed toxicological assessment procedures, still less ecotoxicological procedures. Moreover, the sheer number and scope of the tests required appear to us to be potentially overwhelming. These problems are explored further below and also raise important questions about regulatory procedures, discussed in detail in Chapter 4.

---

ii   A breed of rabbit suffering from a rare genetic defect resulting in fatally high levels of cholesterol in the blood. This strain has proven invaluable in researching human cholesterol-related cardiovascular disease.

3.89 The conventional approach in toxicology assumes that the amount (mass) of toxic material present relates directly to the severity of harm. Testing materials to determine toxicity involves exposing test organisms to increasing amounts of a substance to identify the doses (or more correctly, the exposure concentrations) at which different effects occur (e.g. sublethal effects on growth and reproduction, or lethality).[124] Safe levels of exposure concentration are normally estimated by extrapolating from test results to concentrations below those at which harm can be detected (the predicted no effects concentration or PNEC). These doses or concentrations are usually expressed in terms of a substance per unit volume. This is adequate for most materials but may be inappropriate at least for some nanomaterials because the chemical activity, and hence the toxicological activity, of nanomaterials are often restricted to the surface layer. For any given weight of material the surface layer, and as a consequence the amount of chemical activity per unit weight, increases dramatically as the size of the particle decreases (see table 2.1).

3.90 The smaller the particle the greater the surface area and the greater the potential toxic effect for a given weight of material. The current testing protocols, both at EU level through REACH (Chapter 4) and internationally, will have to be revised to take account of this factor. At the start of this chapter we noted that the European Commission has sought advice on how to evaluate nanomaterials from one of its advisory committees, SCENIHR, and internationally through the OECD. It is important that this work reaches a conclusion quickly so that those responsible for assessing the safety of nanomaterials have clear guidelines to follow when testing.

3.91 Nevertheless, even if new and more effective risk assessment procedures are developed over the next 2-3 years (probably an optimistic timescale), it will be several more years, possibly decades, before the toxicology and ecotoxicology of significant numbers of nanomaterials can be properly evaluated (figure 3-I). The problem is compounded by capacity restraints on testing facilities and expertise and the extra demands already faced for testing to implement the REACH regulations. The measures that can be put in place in the interim to reduce risks to people and the environment are discussed in Chapter 4.

3.92 Beyond problems associated with designing and carrying out adequate ecotoxicology and toxicology tests in a timely manner, there is the difficulty of obtaining a consistent quality standard of nanomaterials to test.[125] The composition and proportion of nanomaterial provided by suppliers can vary between batches and even within a batch as material ages over time. Oxidation or physical processes, such as agglomeration, can significantly alter the properties of a material, as can the incidence of light, leading to different results when materials are tested under different light conditions. In many toxicological studies of manufactured nanomaterials the experimental material itself is not adequately characterised in terms of particle size distribution or other important physico-chemical properties. This makes it very difficult to interpret experimental data to be sure that experiments are testing the same substance and to calibrate for dosage effects. We have been told that it is not unknown for some trials to show no effects and others, on allegedly the same material, to report an impact.[126]

3.93 Most manufactured nanoparticles consist of a combination of constituent parts (see Chapter 2), each with the potential to affect overall behaviour in organisms and the environment. Consequently, it will not suffice to design tests that focus solely on the nanoparticle core. Experiments need to account for the potential impact that any one of the constituent parts may have on the overall health of living organisms. It is also necessary to isolate the toxic effects of the nanomaterial itself while eliminating the confounding influence of other substances, such as the media used to disperse the test substance (3.28). This on its own adds considerably to the number of tests required to understand how an individual nanomaterial behaves in organisms and the environment.

3.94   Once released to the open environment or to a living organism, the constituent parts of a manufactured nanomaterial and the 'core' material itself are likely to alter with time, e.g. the capping material might conceivably degrade to expose the central core. This adds yet another tier of potential health and environmental risks to consider in testing.

3.95   Adding hugely to the complexity of ecotoxicological testing is that many characteristics of the environment can change the availability, mobility and toxicology of manufactured nanoparticles, e.g. pH, salinity, the presence or absence of organic matter. At the moment we have little or no idea of how important these complex effects might be, but the work required to find out is enormous and will take many years to complete.

3.96   Aggregation, along with the size, morphology and kinetics of the aggregate, is one of the key determining factors for the bioaccumulation and ecotoxicity of nanoparticles.[127] As a particle mass grows in size through aggregation it has generally been assumed that the toxicity of a nanoparticle will diminish. However, we have received evidence that the process of aggregation does not necessarily constitute a permanent change. Other processes such as re-suspension and disaggregation may result in a reversal of the aggregation.

3.97   In laboratory tests the degree of aggregation is very difficult to control as various standard experimental practices may induce or discourage the process. The extent of aggregation may consequently differ from that expected in the natural world. This has consequences for the dose–response curve, which will reflect the toxicity suggested by an artificially produced degree of aggregation and may invalidate the standard PNEC as a result.

3.98   Confronted with the enormity of the tasks ahead, we know that work is ongoing both in the UK[128] and the US[129] to develop algorithms to help identify the manufactured nanomaterials of greatest concern. This work needs to be taken forward as a matter of urgency. Once such studies have reached an advanced state it should be possible to develop predictive models of toxicological impacts, as has been done for toxicology more generally (e.g. QSARs, 3.69). But this remains a very long-term goal.

3.99   An alternative way to deal with the scale of the task confronting toxicologists and ecotoxicologists is to note how many materials can feasibly be tested under current toxicity testing regimes. So far, only about 3,000 of the 30,000 bulk chemicals in common use in the EU have been formally assessed for health and environmental effects, although it is likely that more chemicals have been tested by their manufacturers who will possess some knowledge concerning their physico-chemical characteristics. Unless there are orders of magnitude increases in efforts to test new nanomaterials coming onto the market, it will be many years before toxicity test data become available for the manufactured nanomaterials that are currently in use or which are under development.[130] Nonetheless, we believe the task must be attempted. As discussed earlier in this chapter, a good start would be to develop laboratory-based test methods that examine the ability of manufactured nanomaterials to be handled by the physiological and cellular defence systems. This might at least provide a basis for prioritising nanomaterials that behave abnormally for further toxicological evaluation.

3.100 There have already been attempts to define priorities for toxicity testing and development of appropriate protocols. In a report of the findings from a nanotoxicology workshop held in April 2006 at the Woodrow Wilson Center[131] the following conclusions were drawn:

- for all types of toxicology tests, the best measures of nanoparticle dose needed to be determined;

- a standard set of nanoparticles should be validated by laboratories worldwide and made available for benchmarking tests of other newly created nanoparticles (3.8);

- a battery of tests should be developed to uncover particularly hazardous properties using a tiered approach; and

- in the long term, research should be aimed at developing a mechanistic understanding of the numerous characteristics that influence nanoparticle toxicity. Predicting the potential toxicity of emerging nanoparticles (3.98) will require hypothesis-driven research that elucidates how physico-chemical parameters influence toxic effects on biological systems.

Although this work was undertaken a few years ago, progress towards these goals is still painfully slow.

3.101 At the NATO Advanced Research Workshop held in April 2008 in Portugal, extensive discussions resulted in proposals for the augmentation of current risk assessment procedures for nanomaterials. This included development of additional toxicity tests using a wider range of species representing more phyla and paying more attention to the likely fate of nanomaterials when selecting test species. For example, as nanoparticles often aggregate and accumulate in sediment, it might be more appropriate to conduct some tests with deposit-feeding organisms rather than pelagic fish species. It was also noted that biochemical toxicity might not be the only mechanism by which ecological effects are generated by nanoparticles. Recent research has demonstrated behavioural changes in annelid worms encountering low concentrations of aluminium oxide nanoparticles in sediment.[132] If chemosensory detection is involved this might have important consequences for finding food, detecting a mate using chemosensory systems or the detection of chemical clues by larvae.

3.102 Risk assessments based on the aforementioned toxicity tests and other information need to be carried out on nanomaterials at each stage in their life cycle prior to widespread use.

3.103 The previous sections have identified the plausibility of adverse biological effects. We were also mindful that there might be non-biological effects of manufactured nanomaterials on the environment and so requested evidence for consideration. During the course of the study no evidence of this nature has been brought to our attention.

ENVIRONMENTAL RECONNAISSANCE AND SURVEILLANCE

3.104 To assess risks posed by manufactured nanomaterials it is necessary to gain an understanding of the potential level of exposure to them in the natural environment. Exposures from air, water, soil or sediment are all plausible, as is exposure via food. The desk top assessment by the Central Science Laboratory (3.31) is encouraging but needs validation from field measurements. The conventional approach for more familiar potential pollutants involves environmental monitoring. Numerous researchers have highlighted to us the need to track movements of nanomaterials through the environment and the lack of the means to do so at present.[133] However, as we noted earlier, although the currently available monitoring instruments for some (but by no means all) manufactured nanomaterials may possess a high level of precision and accuracy, they are cumbersome and are currently not widely available. At present, the most appropriate instruments are not robust enough and the methods are too time consuming to permit routine analysis of large numbers of different kinds of environmental samples for the presence of even a limited range of manufactured nanomaterials.

3.105 Techniques will need to be extremely sensitive and able to distinguish different physico-chemical forms of nanomaterials, usually against a background of natural nanoparticles with a similar structure and chemistry. The most promising approaches involve advanced separation, spectroscopic and microscopic methods such as flow field-flow fractionation inductively coupled plasma mass spectrometry (FlFFF-ICP-MS)[134] and two-dimensional field-flow fractionation-liquid chromatography.[135]

3.106 Assuming that the technological challenges of appropriate instrumentation can be overcome, it will be necessary to determine the amounts, forms and bioavailability of different nanomaterials in the environment and in organisms, for both the material itself and for any derivatives. This latter point is important as some secondary products could turn out to be more harmful than their parent material.

3.107 Improvements in current systems for environmental monitoring would increase the chance of detecting the early warning signs of damage to populations and communities of organisms, and of changes in ecosystem structure and function. This approach focuses on detecting effects then tracing causes, rather than drawing on *a priori* assumptions. But we do not underestimate the challenge of setting up appropriate biomarker/bio-indicator monitoring systems.

3.108 Among the many challenges is that current concentrations of manufactured nanoparticles present in the open environment are likely to be small when compared with those of naturally-occurring and 'derived' nanoparticles (3.24). Consequently, monitoring programmes will need to be capable of detecting low concentrations of manufactured nanoparticles against a relatively high background concentration of naturally-occurring materials and the potential lability of some nanoforms under the test conditions. Given that the toxicology of nanomaterials may derive as much from the number of particles as from their bulk concentration, this challenge will need to be met. Based upon the current level of investment in nanotechnologies (see Chapter 2) it is likely that, for many years to come, nanomaterials will be released, intentionally or unintentionally, into the natural environment and it will be important to determine where they accumulate and for how long they persist. Therefore, it is important to develop a hierarchy of manufactured nanomaterials for monitoring those that are likely to be most harmful.

3.109 Another way of augmenting risk assessment is to actively look for sites and organisms at or in which nanomaterials might accumulate, thereby enhancing exposure assessment. Sediments below sewage outfall pipes, river water and sediments downstream from major conurbations, and coastal marine sediments are good examples of sites where semi-permeable membrane devices might be deployed to determine whether nanomaterials can be concentrated and detected more readily (especially in sediments).[136] There is also a need to target species that could act as accumulators such as sediment-feeders and aquatic fungi (3.47).

NANOMATERIALS IN THE FUTURE

3.110 While primarily gathering evidence on first and second generation nanomaterials we have been alerted to the likely development of third and fourth generation nanoproducts (figure 2-V). These materials might involve self-assembly capabilities, self-replication and artificial intelligence. There are suggestions that the newly-emerging discipline of synthetic biology might utilise nanotechnologies and nanomaterials in the pursuit of novel products, some of which may have military and space applications where enhanced performance may outweigh cost factors. Much of the discussion of these products is considered to fall well outside conventional regulation of chemicals; their properties raise wider ethical issues as well as health and environmental ones. NATO's 2008 science programme included several conferences and workshops, for example, examining molecular self-organisation (8-12 June, Kyiv, Ukraine), and the environmental and biological risks of nanobiotechnology, nanobionics and hybrid organic-silicon nanodevices (18-20 June, St Petersburg, Russia).[iii]

3.111 We are of the view that it is not too soon to consider the challenges that later generation nanomaterials will pose to conventional procedures for evaluating their potential threats or to the measures which might be incorporated into their design to avoid or minimise such threats. While these challenges lie outside the scope of this study, the governance issues discussed in Chapter 4, in particular how society deals with novel technologies in the face of profound uncertainties, apply in principle to any novel material, and may be helpful as we struggle with the benefits and risks posed by third and fourth generation nanomaterials.

3.112 Even for first and second generation nanomaterials it is reasonable to assume that certainty about the potential risks posed by manufactured nanoparticles may not be achieved for many years and some risks may never be perceived unless or until significant environmental impacts are discovered and traced back to a particular exposure. The sheer complexity of these issues means that we have to try to develop rapid, highly-parallelised, miniaturised screening tests, in an attempt to identify the key substances and conditions for further study.

3.113 We recognise that a great deal of effort has gone into research into the implications of manufactured nanomaterials in terms of their impact on health and the environment. There has also been co-ordination of efforts at OECD level and within the EU (3.3-3.10). Nevertheless, this effort is disproportionately small relative to investment in developing new nanomaterials. In the US, investment in this area is assessed at around 3.5% of total expenditure on nanomaterial research and development.[137] In the UK, the Department for Innovation, Universities and Skills (DIUS) was unable to provide data to enable us to make an equivalent calculation because the baseline periods for expenditure on implications and total research were not comparable, but the proportion is again clearly low.[138] This needs to change, though it is unreasonable to expect any one country to shoulder the burden alone.

3.114 This research needs to be undertaken on a more systematic and strategic basis, which is difficult to deliver under response mode funding as currently used in the UK as the main driver. We appreciate that the Government did not wish to take up the recommendations of the Royal Society and Royal Academy of Engineering report for a new research centre, but **we strongly recommend a more directed, more co-ordinated and larger response led by the Research Councils to address the critical research needs raised by this report, with emphasis on regulatory and policy programmes.**

iii   More information is available at: www.nato.int/science

3.115 These needs include:

- The validation of *in vitro* tests against *in vivo* models.

- Evaluation of methodologies for predicting the likely fate and effects of nanomaterials based on their physical and chemical properties as well as their novel properties and, where possible, the development of exposure scenarios.

- Based on the significant gap in our knowledge, the programme of directed research should ensure a concerted and co-ordinated effort is made to better understand the principles that drive the toxicity of manufactured nanomaterials and how individual properties interact to enhance or diminish toxicity profiles both *in vitro* and *in vivo* with a long-term objective of developing predictive toxicology.

- The enhancement of *in situ* monitoring and surveillance methods to provide early warnings of unexpected effects of novel materials and to permit timely remedial action.

- The research programme should pave the way for much greater interdisciplinary co-operation, including co-operation between those engaged in medical toxicology and those in ecotoxicology, so as to enhance the development of robust test systems and also to act as a catalyst for early warnings from observations on lower organisms to be extrapolated to humans.

3.116 We are also convinced that the research capacity to deliver the necessary volumes of work to address these issues, when combined with other challenges such as meeting the needs of REACH for other chemicals to be tested, is likely to overstretch the existing research community, particularly in the area of toxicology.

3.117 **We recommend that urgent attention is given to undergraduate and postgraduate training in toxicology across all of its domains and that DIUS, the university sector and the professional societies that represent medical toxicologists and ecotoxicologists establish new initiatives to build multidisciplinary capacity in this field.**

3.118 Even with such a programme of directed research and increases in capacity we believe that it will inevitably take a long time to address the need for data to underpin regulation. This is compounded by the fact that we face the challenge of further cycles of innovation and the introduction of new materials and ever more demanding issues to address with more complex nanomaterials.

Conclusions

3.119 From an extensive review of the original published literature that, in nanoscience, has been especially prolific over the last four years, several important conclusions can be drawn:

- There appears to be little consensus over the critical or even most important characteristics of manufactured nanomaterials that drive their toxicity profiles.

- Little information is available on how the various physical and chemical properties interact to generate an overall toxicity profile for a particular nanomaterial.

- There has been little attempt to use standard particles to study individual characteristics and their interactions, nor concerted attempts to develop indices similar to quantitative structure–activity relationships (QSARs) that are currently being used for chemicals.

- Knowledge on the medical applications of nanomaterials with respect to organ, cell and subcellular localisation should be harnessed to aid understanding of predictive toxicology.

3.120 The number of experimental centres involved in nanotoxicology is small and they seem to use different materials and experimental protocols. There is an urgent need for standardisation and co-ordination of research effort and focus in this field. There is remarkably little link between knowledge gained from ecotoxicology and that from the study of toxicity in higher organisms including humans. Greater co-ordination and application of basic principles is needed between the two activities.

3.121 The integrative and multidisciplinary nature of toxicology as a science and a profession requires special skills that, over the last twenty years, have been declining on account of reduced training and career development dedicated to this subject. As we have discussed, the demands for high-quality science, the integration needed between the biological and physical sciences and the urgent requirement for scientists to integrate findings from animal toxicology and ecotoxicology demands that more attention is given to toxicology training in our higher education institutes to increase the cadre of properly trained individuals needed to take on the challenges of nanotoxicology.

3.122 In this chapter we have established that there is a plausible basis for concern that some manufactured nanomaterials could present a hazard to human health or the environment. We have reviewed the research effort to address these concerns and have made recommendations to augment that programme and to put it on a more systematic footing. We have concluded that there remains a great deal to be done and that it will take a very long time to address these questions.

3.123 However good the research effort, significant uncertainties and areas of ignorance will remain. Effective monitoring of the environment is therefore crucial to give early warning of unexpected impact. In Chapter 4 we examine the current governance framework for chemicals and consider how this addresses the challenges of manufactured nanomaterials. We also stand back from our focus on nanomaterials elsewhere in the report and draw on our analysis of this sector to recommend ways in which future new technological developments should be governed, taking into account the global context and the rapid pace of introduction of novel and new materials of all kinds.

Chapter 4

THE CHALLENGES OF DESIGNING AN EFFECTIVE GOVERNANCE FRAMEWORK

INTRODUCTION

4.1 We concluded in Chapter 1 of this report with the recognition that the development and widespread application of novel materials present society with an instance of the 'control dilemma'. In the early stages of development of a technology we do not know enough about its future implications to establish the most appropriate management regime. But by the time problems emerge, the technology is likely to have become too embedded to change without significant social or economic disruption. In considering the potential for novel materials (exemplified in this report by nanomaterials) to enter organisms and the environment in ways that could be damaging, we recognise that we are operating under conditions of partial knowledge and significant ignorance.

4.2 The challenge of controlling novel materials is exacerbated by the fact that they are seldom encountered as discrete entities, but are likely to be contained within products. This introduces two further implications, making them very difficult to control. First, it means that they will not necessarily be recognisable and may, therefore, escape regulatory attention. Second, in the context of a globalised economy and world trade, they are likely to become ubiquitous. Controls established in one country or region may not be observed by producers of goods which are likely to be circulated worldwide. We refer to this as the 'condition of ubiquity'.

4.3 The conditions of ignorance and ubiquity define the control dilemma for novel materials. Focusing on the instance of manufactured nanomaterials, Chapters 2 and 3 have shown how the novel properties and functionalities of these materials pose challenges for those charged with protecting the public and the environment from harm. This chapter considers the challenges of regulation and of governance in a wider sense. It proposes ways to manage the control dilemma for nanomaterials and, by extension, for a wider class of novel materials.

THE CHALLENGES PRESENTED BY NANOMATERIALS

4.4 Nanotechnologies cover an enormous range of possibilities with profound implications. Looking to the future, what are sometimes characterised as third and fourth generation nanotechnologies (Chapter 2) raise ethical and political questions (concerning, for example, human identity, performance and privacy) which call for the widest possible debate.[1] We emphasise, as we did in a much earlier discussion of nuclear technologies,[2] that our concern should not only be with the position at present, or even in the next decade, but with what it might become within the next fifty years.

4.5 Even first and second generation nanotechnologies – those available now or likely to become so in the near future – raise questions about ownership, control, the direction of innovation and the distribution of gains and losses.[3] Such questions embrace, but extend well beyond, concerns about potential risks to human health and the environment. Evidence from many sources has convinced us that even addressing the health and environmental risks will be a challenging task. Hence, while recognising the significance of bigger questions (to which we return in the final section of Chapter 4), we have focused in the body of this report on the health and environmental risks that might be presented by manufactured nanomaterials, and on ways in which they might be regulated and governed.

4.6   Our extensive enquiries produced no evidence of actual harm, either to human health or to the environment, which could be attributed to manufactured nanoparticles (Chapter 3). But this absence of evidence is not conclusive: as noted in Chapter 1, there are numerous well-documented cases where the unpredicted, harmful effects of new products have not immediately become apparent.[4] It is conceivable that free nanoparticles will behave in quite unexpected ways in the environment or in living organisms, including humans, as discussed in Chapter 3.

4.7   Where plausible pathways and consequences for health or the environment can be identified, the rapid and widespread introduction of materials with novel properties and functionalities demands at least that we be vigilant, even where there is no specific evidence of damage. Often the plausibility of harm centres on the novel properties and behaviours that provide the rationale for manufacturing such materials in the first place.

4.8   In a few cases, emerging evidence suggests that certain types of nanomaterial have the potential to pose significant risks to the environment or to human health (Chapter 3). For example, carbon nanotubes of a particular length are reported, under experimental conditions, to cause changes similar to those caused by asbestos,[5] and ionic silver, unlike bulk silver, may be toxic to living organisms such as bacteria and fish.[6, 7]

4.9   Obviously, research designed to improve our understanding of the behaviour of such materials in the human body and the ambient environment is highly desirable. We acknowledge the impressive efforts currently being made internationally, not least through the Organisation for Economic Co-operation and Development (OECD), to assess the human health and environmental implications of nanomaterials. Underpinning this work are major programmes designed to rigorously characterise nanomaterials, and studies to develop appropriate toxicity and ecotoxicity tests. Such research promises to produce new knowledge and reduce uncertainties, but, as we point out in Chapter 3, this may well be a lengthy process, it will always be incomplete and it will not necessarily deliver results before irreparable harm is done to individuals or ecosystems.

4.10   An important question therefore is whether existing regulatory frameworks provide sufficient safeguards for human health and the environment, given the predicted rapid growth in the number and availability of manufactured nanomaterials (Chapter 2). In this chapter, we examine current regulatory regimes, primarily in the UK and Europe, assess their adequacy in this respect, and consider whether we need incremental change and adaptation, or radical alternatives. We then go on to consider the role of regulation within the broader context of governance.

4.11   To be effective and worthy of public trust, any governance system must be able to demonstrate that it has the technical competence to understand and manage the systems for which it is responsible. It must also be inclusive and capable of demonstrating fiduciary responsibility towards its constituents.[8] Effective and trustworthy governance arrangements must therefore have at least four key qualities. They must be *informed, transparent, prospective* and *adaptive*. To achieve these characteristics they also need to be supported by skilled regulatory bodies and decision-making processes that deliver proportionate outcomes.

4.12   Effective governance requires more than just top-down regulation. The uncertainty and ignorance that characterise our lack of understanding of the impacts of nanomaterials mean that traditional regulatory mechanisms on their own may not provide protection without adversely affecting innovation. We are likely to have to adopt a wide suite of measures and involve many actors. The process will be characterised by contestation as well as co-operation. Because many of the issues are, as we described in Chapter 1, trans-scientific in nature, it is essential to recognise that the process will be profoundly and legitimately political and cultural as well as technical and economic.

4.13   Unsurprisingly, approaches to the regulation of nanomaterials vary around the world, in part due to the dominance of different world views (Chapter 1), although they also exhibit some similarities. In the European Union (EU), the USA and Japan, for example, the placement of new chemical substances on the market is controlled by legislation designed to protect human beings and the environment, and in all three regions, a notification procedure is required.

4.14   Officials in Japan are confident that their existing regulatory regime, the Chemical Substances Control Law, will adequately manage nanomaterials.[9] This requires the provision of toxicological and ecotoxicological data concerning biodegradation and bioaccumulation, followed by a designation procedure that determines whether additional data are needed. If the rate of biodegradation is low and that of bioaccumulation is high, long-term human toxicity tests must be conducted to obtain information on the risk to human health.

4.15   In the United States, the introduction of new chemicals is controlled by the Environmental Protection Agency (EPA) through the Toxic Substances Control Act (TSCA). The EPA can solicit new tests and ban the manufacture or import of a highly hazardous substance. It is also responsible for an inventory of commercial substances imported to, or produced in, the US. The information submitted includes all the available data on the chemical's identity, production volume, by-products, use, discharge into the environment and disposal practices, and an estimation of human exposure. Additional information may also be requested by the authorities.

4.16   Officials in the United States acknowledge that their regulatory framework may have gaps with regards to nanomaterials, but they believe that the combination of federal and state regulation and multiple overlapping Federal Agency responsibilities, combined with the tort law system, provides a distributed regulatory and early warning system that will pick up any problems that could arise.[10]

4.17   The places and roles of risk analysis and regulation, expert judgement and world views in the governance of new technologies are themselves functions of world views (1.33).[11] Using the variables of 'systems uncertainty' and 'decision stakes', philosophers of science Silvio Funtowicz and Jerome Ravetz have identified three distinctive sets of conditions requiring different governance approaches (figure 4-I). Where technologies are well understood and the consequences of errors are minor from a societal perspective, they argue that governance by rules and regulations based on standardised technical analyses is appropriate. However, as either uncertainty or the impact of error increases, rules-based regulation needs to be supplemented by craft-skills and expert judgements. Where either dimension is high, then decision making is inevitably informed chiefly by world view.

FIGURE 4-I
Three kinds of assessment for decision making[12]

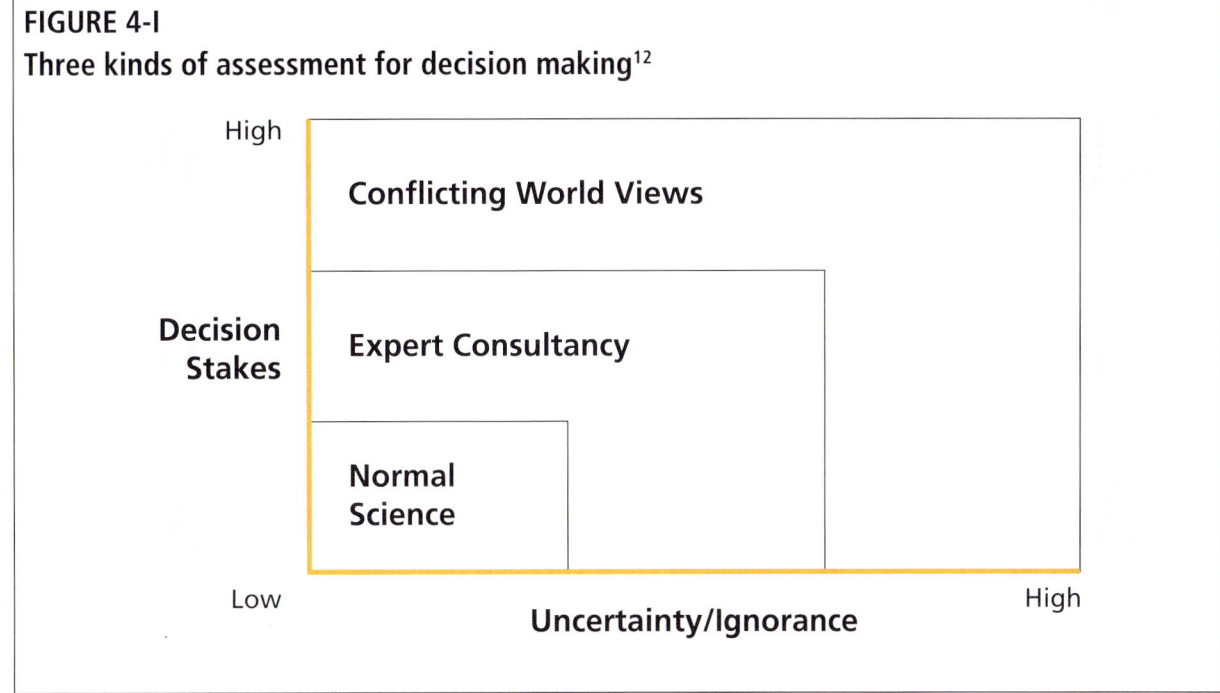

4.18 Of course, where we locate any particular technology in this schema is itself likely to be influenced by our world view. (Generally, regulators try to squeeze new issues into existing rules and routines while environmental campaigners are likely to look for big uncertainties and broad ethical implications.) Nevertheless, the device helps us to think through the extent to which nanomaterials and other novel materials can adequately be regulated through established rules and analytical techniques, the extent to which these may need to be modified in the light of expert judgement about the potential for, as yet, unrealised damage (and the research necessary to resolve these concerns) and, finally, the role of wider publics in shaping the application and use of nano- and other novel materials where uncertainties or potential consequences give rise to concern.

4.19 Hence, the remainder of this chapter follows this three-stage logic, moving from assessment of the adequacy of existing rules and regulations, through a consideration of how they might be modified or extended, to exploration of what additional governance arrangements, if any, would be appropriate. The next section outlines the existing structure of regulation applicable to novel materials in the UK and Europe. We go on to focus on the challenges presented by nanomaterials and their likely widespread deployment, and we consider ways in which regulatory structures might be modified or extended to accommodate concerns about potential exposure pathways and possible harm to the environment and human health. We then consider a number of possible measures that might supplement or reinforce existing regulations, especially during the period when the latter are being adapted to incorporate nanomaterials. Finally, we return to wider societal concerns and how they might be addressed in a governance framework for novel materials and, indeed, for science-based technological developments more generally.

The reach of existing regulations in Europe and the UK

4.20 There are no specific regulations for nanotechnologies or nanomaterials in Europe or the UK. Instead, the manufacture, use and disposal of nanomaterials are covered, at least in principle, by a complex set of existing regulatory regimes (appendix J). These include the Integrated Pollution Prevention and Control (IPPC) Directive[i] and other consumer and environmental protection regimes. In addition, particular product types are covered by a series of 'vertical' regulations, including REACH, which is concerned with the **R**egistration, **E**valuation, **A**uthorisation and Restriction of **Ch**emical substances,[13] and product- or sector-specific regulations for pharmaceuticals, veterinary medicines, pesticides and biocides. There are also specific regimes dealing with toys, cosmetics and end-of-life practices, such as the Waste Electrical and Electronic Equipment (WEEE) Directive.[14]

4.21 The combined effect of this legislation is to impose a responsibility on those who manufacture and sell the products to which it applies, requiring them to identify and understand potential threats to human health and the environment, and to minimise or eliminate the risk of adverse effects. In principle, these requirements apply to nanomaterials in the same way as to other substances and products.

4.22 REACH, which came into force on 1 June 2007, transforms chemicals regulation in the European Union[15] and constitutes the regulatory framework of greatest relevance to the governance of novel materials. It regulates existing (already marketed) and new chemicals, covers substances in their own right as well as those in manufactured products and preparations, and applies to imports as well as to substances manufactured in the EU. It is estimated that some 30,000 substances will be covered by this regime.

4.23 Although REACH was designed to provide comprehensive regulation of chemicals and not formulated specifically to deal with nano- or other novel materials, most materials would fall under its existing rules if manufactured or used in sufficient quantities. There is potential for revising the thresholds below which REACH does not currently apply. We return to this issue below.

4.24 REACH operates on the premise of 'no data, no market'. Chemical substances manufactured or imported at or above a threshold of 1 tonne per annum per manufacturer or importer are subject to a registration requirement. This took effect for new substances on 1 June 2008 and will take effect for existing substances during a variable phase-in period ending in 2018, providing substances are pre-registered between 1 June and 30 November 2008.

4.25 This registration obligation imposes a duty on industry to provide data on the substances concerned, including physico-chemical, toxicological and ecotoxicological data. The volume of data required depends upon the volume of production and to a lesser extent upon the level of risk posed by the substance concerned. The ten tonne threshold is a key one, triggering as it does the obligation to undertake a Chemical Safety Assessment and submit a documented Chemical Safety Report.

i   The IPPC Directive is in the process of being replaced by a directive on industrial emissions. The proposals can be viewed at: http://ec.europa.eu/environment/air/pollutants/stationary/ippc/proposal.htm

4.26   Registration documents will be submitted to the newly-established European Chemicals Agency, which will check the completeness of the dossiers submitted, and conduct a more detailed evaluation of a sample of these. The Agency will draw up a list of priority chemical substances, for which Member States will be required to conduct a more detailed evaluation of hazards and risks though their national competent authorities. In the UK, the national competent authority is the Health and Safety Executive (HSE).

4.27   In addition to the information requirements associated with registration, REACH provides two primary mechanisms to regulate market access for certain chemical substances. First, it contains a revisable list of 'substances subject to restrictions'. The content of this list will be revised where there is evidence that any given substance poses an unacceptable risk to human health or to the environment. Second, certain 'substances of very high concern', placed on a priority list for action, will require 'pre-market authorisation' from the competent authorities even if they are manufactured or imported at very low volumes (below the one tonne threshold for registration).

4.28   Certain substances are automatically viewed as substances of very high concern and as being open to inclusion on the list of substances requiring authorisation. These include category 1 or 2 carcinogens, mutagens or reproductive toxicants, as well as those which are persistent, bioaccumulative and toxic (or very persistent and very bioaccumulative) and which meet criteria set out in Annex XII of the Regulation. Certain other substances may also be made subject to this authorisation requirement. These include, for example, endocrine disruptors, where there is scientific evidence of probable serious effects on human health or the environment giving rise to an equivalent level of concern (equivalent to the level of concern raised by those substances automatically viewed as substances of very high concern). The process for selecting those substances requiring authorisation will be in two stages, with both a candidate and a final list being drawn up. Both the European Commission and the Member States, as well as the new Chemicals Agency, will be closely involved in this process. Interested parties enjoy an opportunity to make comments in the course of the adoption of both the candidate and final list, and the European Parliament is involved in the adoption of the final list.

4.29   For certain substances,[ii] authorisation will only be granted where there is no suitable alternative substance or technology, suitability being assessed with reference to overall risks and with regard to the technical and economic feasibility of proposed alternatives. REACH also gives the European Chemicals Agency powers to re-assess previously authorised chemicals on the basis of new data, and to restrict or ban their use if evidence of harm is sufficiently compelling.

4.30   REACH places considerable emphasis on transparency and access to information. It does so by way of a variety of mechanisms. First, it regulates the provision of information throughout the chemicals supply chain, requiring the continued provision of a Safety Data Sheet for substances with specified properties. It further integrates the classification and labelling of chemicals with the current Chemicals (Hazard Information and Packaging for Supply) Regulations 2002 (CHIP) in an effort to ensure a high level of protection for consumers and workers. Finally, REACH contains elaborate provisions on access to information, only permitting exclusion on a routine basis of certain commercially confidential sensitive information.

ii   See Article 60(3) of the Regulation for a description of those substances to which this substitution analysis will apply.

4.31 At least in principle, it would appear that REACH is capable of meeting the criteria for effective governance, or at least its regulatory component, outlined earlier in this chapter (4.11). Decision making is intended to be well *informed*, due to industry responsibility for the provision of information and the role of the new Chemicals Agency and its expert advisory committees. The system is intended to be *transparent* and *open*, with multiple opportunities for interested parties to participate, and an emphasis on the collation and dissemination of information on chemical substances. It seeks to be *prospective* or *forward-looking*, in that REACH applies to both new and existing chemicals. And finally, REACH is intended to be *adaptive* and *flexible*. It provides a framework for the continuing review of authorisations, and even for the revision of key elements of the regulation itself.

4.32 In addition, REACH provides opportunities for Member States and the European Commission (and to a lesser extent the European Chemicals Agency) to take the initiative in shifting the regulatory agenda by highlighting new challenges and requiring a common EU-wide response. For example, Member States and/or the European Commission are empowered to initiate a new restrictions process or a process for adding to the list of substances requiring authorisation.

4.33 Important exclusions from REACH include substances which are covered by alternative regulatory regimes, such as human and animal medicines, foodstuffs, animal feed, cosmetics and medical devices. Also excluded are certain substances considered to present minimal risk[16] because they are so widespread and well understood as to be beyond the need for further evaluation. At present these include water, carbon dioxide and nitrogen. Interestingly, carbon and graphite were dropped from the list of exempt substances in June 2008 because of concerns about their nanoforms.[17] (Other carbon forms, diamond and carbon black, were not on the exempt list.)

4.34 In general, the product- and specific-sector regulations cover the same ground as REACH and provide broadly equivalent safeguards. One exception is the Cosmetics Directive, which is distinctive in explicitly prohibiting animal testing after 2009, as discussed in Chapter 3.

EXTENDING OUR REACH

4.35 Regulatory instruments like REACH have not been designed with nanomaterial products and their applications in mind, so it is a matter for concern that their risks might not be captured effectively within the current framework. A number of significant gaps in existing regulations have been identified in studies commissioned by the UK Government and other administrations.[18, 19] A question that follows is whether existing regulations can be modified to close or narrow such gaps, or whether a regime specific to nanomaterials, or nanotechnologies more broadly, would be preferable.

4.36 One problem is that some nanomaterials may simply escape attention. Under REACH, nanoscale versions of existing substances (e.g. titanium dioxide) are treated in the same way as the equivalent bulk material, even if they have very different properties (and indeed are being manufactured for this reason).

4.37 However, the most significant potential limitation affecting the application of REACH to nanomaterials is the one tonne threshold for registration. Because of the very large number of (often highly interactive) particles present even in tiny quantities of a nanomaterial (see table 2.1), one tonne may be too high a threshold to capture potentially problematic effects.

4.38   The authorisation and restriction procedures outlined above (4.24-4.32) are applicable irrespective of production volume, and might therefore capture some of the potential impacts of nanomaterials. But for these procedures to come into play there would have to be evidence that a particular class of nanomaterial posed an unacceptable risk to human health or to the environment. If such evidence emerges, the competent authorities in the EU have powers under REACH to greatly restrict or to ban the use of those materials (4.29). While such powers might be sufficient in principle, in practice the prediction of harm is problematic and techniques for monitoring nanoparticles in organisms and the environment are either non-existent or embryonic (Chapter 3).

4.39   Current regulations may also have consequences for the generation of new data on risk.[20] For example, if a substance is classified as hazardous, REACH requires the supplier to provide further information on the nature of the hazard and the possible risks involved (4.25). However, if a material is not initially classed as hazardous for a particular use (or uses), and this non-hazardous classification is extended to a nanoform of the substance, then there will be no requirement to generate data about the specific effects of the nanoform (for example on its fate in the environment, organisms or people). A nanomaterial might then move through its entire life cycle (2.54) without any requirement for further assessment of its properties, despite the possibility that, although it is not considered harmful to human health or the environment in its approved use, it might have the capacity for adverse impacts at some other stage, for example, as a result of release of the products of abrasion or combustion.

4.40   REACH places ultimate responsibility for the safety of products on the manufacturer or importer, rather than on the regulator. We are aware that a number of major companies are going to great lengths to try to assess the possible risks posed by nanomaterials under their control. However, they might not be carrying out the tests that would reveal unexpected hazards, and the right information may not be asked for by the regulator. To some extent, these issues apply to the risk assessment of any substance, but for reasons that we will come to below (4.53), they are particularly challenging in the case of nanomaterials. We doubt that smaller companies and importers have either the capacity or the resources to deal with these difficult issues.

4.41   On the issue of whether modified regulations or new (and possibly dedicated) arrangements are required to cover nanomaterials, the European Commission has opted for the former. It does not regard specific regulation on nanomaterials as feasible in the European context because of the difficulty in establishing links between very different pieces of legislation and the need to negotiate internationally to establish a regulatory process.[21] Therefore, it has adopted an incremental approach, intending to adapt existing laws to the regulation of nanotechnologies.

4.42   The position of the UK Government is less clear: in a recent statement it has suggested that further evidence is required to help determine whether new or amended legislation will be necessary, at least as far as free engineered nanoscale materials are concerned.[22]

4.43   'Regulatory gap' analyses have tended to conclude that the existing framework is capable of adaptation to make it fit for purpose in dealing with nanomaterials, providing that the adaptation is underpinned by research to assess impacts and inform the setting of standards.[23]

4.44   After careful consideration, we agree. We have not seen convincing evidence of the need for a special regulatory regime for nanomaterials, let alone for the wider class of materials considered to be novel. Not only is the legislative field already crowded, but nanomaterials do not constitute a unified class of substances (Chapter 2). Most importantly, we have argued that the issue with all materials is their functionality (Chapter 1). It is not the fact that they are created by any particular technology that is important, or even, in the case of nanomaterials, that they are of a particular size. What matters is what they *do*, and the implications of their properties and functionalities for environmental protection and human health. There is no logical reason why size of particle should in itself provide the basis for new regulatory controls.

4.45   **We recommend that in any revisions to existing regulations, the relevant authorities should focus specifically on the properties and functionalities of nanomaterials, rather than size. Since these properties and functionalities will often differ substantially from those of the bulk material, strict chemical equivalence does not preclude the need for a separate risk assessment.** We are concerned, for example, about the numerous kinds of carbon nanoparticles, nanofibres and nanotubes, as well as fullerenes, whose properties are fundamentally different from those of carbon black or graphite, and about nanosilver, which exhibits quite different toxicity to the bulk metallic form.

4.46   We have argued above (4.35-4.40) that REACH embodies many elements of an effective adaptive management system, and we are persuaded that its adaptation (and adaptation of the product- or sector-specific regulations) to cover nanomaterials is feasible in principle. But we are in no doubt that, if this is to be achieved, some substantial modifications will be needed.[24] We welcome the fact that the European Commission has asked its Scientific Committee on Emerging and Newly Identified Health Risks (SCENIHR) for a scientific opinion on how the application of REACH might be modified in this respect.[25] We also welcome the first step made towards taking nanomaterials into account (4.33) by removing carbon and graphite from the list of substances exempt from REACH.[26]

4.47   **We recommend that the UK Government should press the European Commission to proceed with urgency, in consultation with Member States, the European Chemicals Agency and SCENIHR, to review REACH and the product- or sector-specific regulations. The object of the review should be to amend the regulations to facilitate their effective application to nanomaterials and the provision of adequate testing arrangements.**

4.48   **We recommend the establishment of clear priorities for testing, beginning with those nanoparticles with functionality which suggests that they might pose the greatest risk of harm to the environment or to human health.**

4.49   The most plausible risks seem at present to be posed by free manufactured nanoparticles that are deliberately introduced into the environment (e.g. iron nanoparticles used to remediate contaminated land), by materials that must inevitably find their way into the environment (e.g. nanoparticles used in sunscreens, cosmetics, water purification systems and diesel additives), and by materials that will probably find their way into the environment by abrasion and wear-and-tear of some larger consumer items (e.g. carbon nanofibres or nanotubes abraded from car tyres or clothing). Prioritisation should be based on considerations such as these, combined with growing knowledge of the behaviour of different classes of nanomaterials. We learned in Japan, for example, that materials known to degrade slowly and to bioaccumulate are singled out for particularly careful scrutiny (4.14).

4.50   Given the problems experienced in the past with substance-by-substance risk assessment of bulk chemicals in the EU (particularly its slow progress as described in Chapter 3), and the additional complexities of nanomaterials, alternative options are worth considering. For example, regulators may wish to limit human or environmental exposure to freely available manufactured nanoparticles (or those that become free during the life cycle of a product), for example through mandatory design requirements, at least until such time as a risk assessment has reduced uncertainty about possible ill-effects.

4.51   For reasons outlined above (4.37), the weight thresholds that trigger procedures under REACH may need to be reduced in the case of nanomaterials, particularly for those whose properties and functionalities give rise to concern. The European Commission is aware of this issue and acknowledges that it will be necessary to monitor the situation and possibly refine the thresholds over the coming years.[27] **We recommend that, as REACH is adapted to meet the challenges presented by nanomaterials, particular attention should be given to the issue of weight thresholds. In view of the persistent uncertainties involved, a precautionary approach should be adopted when determining new, lower thresholds for nanomaterials.**

4.52   To summarise, so far we have argued that the existing regulatory framework has the potential capacity to manage the possible risks associated with nanomaterials but, as it is currently being implemented, it does not adequately do so. We believe that the most significant regulations are capable of adaptation; and that a strategy of modifying existing arrangements needs to be pursued with some urgency.

BEYOND OUR REACH

4.53   However, we do not underestimate the scale of the challenge involved. Indeed, we remain deeply concerned about the timescales required to negotiate modifications and to gather the necessary data. Much of the basic scientific information is not yet available, and we were told that the goal of accurately predicting the toxicity and environmental behaviour of manufactured nanomaterials is likely to be several decades away.[28] On this point, the issues identified in Chapter 3 bear summarising again:

- The novel properties of nanomaterials mean that standard toxicity tests may be inappropriate for the identification of potential harm, or impossible to carry out on some classes of nanoparticles.

- We know little about the mobility, accumulation, degradation and ultimate fate of most nanomaterials in the environment.

- This means that entirely new toxicity and ecotoxicity tests may need to be developed for some types of nanomaterials and agreed at a European and OECD level. This will not be achieved quickly, if at all.

- To facilitate the development of testing protocols, much work needs to be done on standardising and characterising nanoparticles. Whilst this work has started (for example, through OECD), it will also, inevitably, take time.

- Each variant of a particular class of nanomaterial may display quite different physical and chemical properties. In some cases there are many variants. Carbon, for instance, can be made into nanospheres, nanofibres, nanotubes, nanosheets and fullerenes, which themselves can be made into tubes and fibres. But the multiplication does not stop there. The properties of nanoparticles can vary with size, shape, charge, coatings and surface characteristics, and with the nature of the medium in which they are incorporated.

- The scale of the problem now becomes obvious. We are no longer dealing with the toxicity and ecotoxicity of 'carbon', or any other material in nanoform, but with a factorial experiment in which the possible combination of conditions rapidly becomes daunting. Of course, many of these multiple types of nanoparticle are not yet in commercial production, but as they come onstream they could risk overwhelming our ability to carry out the necessary toxicity and ecotoxicity tests in a timely manner. Furthermore, any given nanomaterial may find application in many different products, some of which will present more of an exposure risk than others.

- While a fundamental breakthrough in our ability to predict the behaviour of nanoparticles could transform the picture, it is most unlikely to be achieved quickly. Nor, ultimately, can any predictive method provide a guarantee against all possible future harm.

4.54　In such circumstances, regulators face a Sisyphean task. Innovation is, or soon will be, driving new products onto the market at rates that are orders of magnitude faster than they can currently hope to manage with the resources at their disposal. We heard from one regulatory body that it was not even considering how to address third and fourth generation nanomaterials because they were fully occupied with those currently at the commercial stage.[29] The magnitude of the task combined with constraints on resources tends to create an attitude of regulatory fatalism.

4.55　One expert likened the challenge of risk governance in this field to that of shouting a warning to the driver of an express train as it thunders past.[30] An understandable reaction in such circumstances is to call for a moratorium on the manufacture and use of nanomaterials; this position, associated with a world view at one end of the spectrum identified in Chapter 1, has been adopted by some environmental and consumer organisations in Europe and the United States.[31]

4.56　We have not been persuaded that a blanket ban would be appropriate; nor do we consider that such an approach could be justified on current evidence. Nanomaterials are already widely marketed and their production is globally distributed: at present we see no prospect for the negotiation of a global moratorium. Furthermore, as Chapter 2 has shown, nanomaterials are extremely diverse, exhibiting a wide variety of properties and functionalities. In many cases, materials in use have known or potential benefits and there is no particular reason to suspect that they will cause harm. We therefore consider that a blanket ban would be neither practicable nor proportionate. Selective moratoria (for which existing regulations make provision) might, however, be an appropriate precautionary measure in particular circumstances.

4.57　Nor, at the other extreme, can we afford to stand back and conclude that this is all too difficult. None of our arguments so far is intended to imply that developing new tests, striving for predictability and working to adapt existing European and national legislation to manage nanomaterials are a waste of time. But the challenges set out above raise two important questions. One is what might be done in the interim, while we wait for a better informed, and better adapted, regulatory framework. We address this question in the following section.

4.58 The other, more profound, question returns us to the fundamental themes of this report. It concerns how we democratise scientific and technological developments in the face of the control dilemma and the conditions of ubiquity and endemic uncertainty (4.1-4.3). It suggests that the very concept of a gap or an 'interim' holds out a false promise of resolution, because the dilemmas that we face are ultimately social and political as well as scientific and technical. We return to these issues in the final part of this chapter.

4.59 The measures recommended above will take time, possibly decades, to develop and implement, as will the accumulation of knowledge on the properties and potential impacts of nanomaterials. In the meantime, it will be vital to make every effort to narrow the gap, and to do so in two senses. One is in improving our understanding of the implications of nanomaterials. The other is in finding ways to anticipate, and as far as possible avoid, harmful effects that would not be captured by current regulations; this is needed in advance of, or in addition to, the kinds of amendments to the regulations outlined above.

4.60 The first, indeed key, requirement to improve understanding is *research*. As demonstrated in Chapter 3, there is an urgent need for research on the implications of nanomaterials, though it is important to recognise that science will not always reduce uncertainties and indeed might raise new questions and open up new areas for exploration. In 2004, the Royal Society and Royal Academy of Engineering recommended research into possible adverse health and environmental impacts, as well as into societal and ethical issues arising from the development of nanotechnologies.[32]

4.61 In partial response to this recommendation, the Department for Environment, Food and Rural Affairs (Defra) established and chairs a Nanotechnology Research Co-ordination Group (NRCG) to establish research priorities and develop links both nationally and internationally to promote dialogue and facilitate the exchange of information. A number of research programmes are also now underway in the UK, including the joint Natural Environment Research Council (NERC), Defra and Environment Agency 'Environmental Nanoscience Initiative', which has committed £2.3 million to date to research projects investigating the fundamental effects of manufactured nanoparticles on the natural environment.

4.62 Other countries, including the United States and Japan, have also established research programmes. At international level, the OECD Working Party on Manufactured Nanomaterials has a programme of work underway to investigate the safety of manufactured nanomaterials.

4.63 In spite of these investments and work programmes, there remains a substantial mismatch of funding between research on the *applications* of nanotechnologies and nanomaterials and investigation of their *implications* for society, the environment and human health (Chapter 3). We find this disturbing, and we have recommended that the UK Government initiate through the Research Councils a directed programme of research with an emphasis on policy and regulatory questions (3.114-3.115).

4.64   Currently, technology-based innovation in the UK is promoted through a single body, the Technology Strategy Board, but there is no equivalent body for managing and regulating the emergent technological developments. In the case of nanomaterials, responsibilities for negotiating and implementing regulations and for funding research are spread across a number of different organisations and government departments, including Defra, the Department for Innovation, Universities and Skills (DIUS), the Department for Business, Enterprise and Regulatory Reform (BERR), HSE, the Environment Agency and the Scottish Environment Protection Agency (SEPA), and the Research Councils. We considered whether it might be desirable for a single body with regulatory responsibilities for novel materials to mirror the innovation support role of the Technology Strategy Board. However, the nature, scope and applications of nanomaterials, let alone the larger class of novel materials, is so broad as to make this impractical. Furthermore, we are persuaded that a somewhat untidy regulatory landscape of agencies with diverse priorities and overlapping responsibilities offers many opportunities to identify potential problems. We received evidence from officials in the United States that, to date, this approach is considered to have worked well there.[33] **Therefore, we recommend that responsible organisations set up structured systems to keep a watching brief on the development of novel materials and to enhance the sharing of information and the opportunities to work together to identify and manage emerging problems.**

4.65   In the course of our investigations, we also considered a number of other possible measures, in addition to the regulatory adjustments advocated in the previous section, aimed primarily at anticipating and avoiding potential harm from nanomaterials. Some of these measures could be applied within a modified form of REACH while others might be additional to these requirements. None of them is without difficulty.

4.66   We acknowledge that there is considerable potential, within a wider system of governance, for self-policing and the development of *codes of conduct*. A number of such codes relating to nanomaterials have been developed, or are in the process of development, including a *Code of Conduct for Responsible Nanosciences and Nanotechnologies Research* recommended by the European Commission,[34] and the Royal Society and partners' *Responsible Nanocode*,[35] for businesses engaged in nanotechnologies. These are quite precautionary in approach, and can raise awareness, improve vigilance and generate a sense of responsibility. However, in our view voluntary codes of conduct are likely to be most effective when they are backed up at appropriate points by 'harder' legal and regulatory measures.

4.67   One additional regulatory possibility that we discussed was that of extending *product 'take-back'* requirements (exemplified in Europe by the WEEE Directive) to products containing nanomaterials. 'Take-back' is intended to prevent or limit the entry of harmful substances into the environment and, in principle, enables the consumer to return a product to a retailer for recycling. If they can be effectively implemented, we support such regulations for novel materials where their functionality suggests that there may be grounds for concern.

4.68   We concluded, however, that in the current global marketplace there is no prospect that a take-back requirement could effectively provide for the return of nanomaterials to their original producers. We do not see how any such scheme covering a wide variety of consumer products containing nanomaterials would be workable, not least because nanomaterials are already incorporated into many products (including clothing, sunscreens and food packaging) for which take-back seems impracticable, if not impossible. Such a scheme, in applying to nanomaterials indiscriminately, would also violate our 'functionality principle'. We did not, therefore, pursue this option in any detail.

4.69   *Labelling* is another possible tool for the management of nanomaterials. Products are usually labelled for one of two reasons: to communicate a known hazard to consumers; or to provide information about a product so that consumers can make an informed choice.[36] We heard contradictory views on this possibility. There are powerful arguments in principle that consumers should be informed, and some may legitimately wish to know whether products contain nanomaterials. (We note that in some instances manufacturers have considered a 'nano' label to be advantageous.) But labelling might also convey the false impression that nanomaterials have uniform properties and is unlikely (at least at this stage) to be able to provide useful information about impacts on health or the environment. At present, we see no reason to recommend product labelling for nanomaterials *per se,* although where the functionality of the material presents a known hazard, it would have to be labelled in accordance with existing regulations, such as the Hazardous Waste (England and Wales) Regulations 2005.[37]

4.70   We also considered, briefly, whether an international convention, in the style of the United Nations (UN) Persistent Organic Pollutants (POPs) Convention, might usefully be set up to regulate nanomaterials. Initially, this idea had some attraction because of the global nature of nanotechnologies, but evidence presented to us did not support such a step, on the grounds that it was too complex and would in practice be unmanageable.[38] Again, it would be unhelpful to treat all manufactured nanomaterials as a single class of substances, and inappropriate to categorise them all as pollutants. We concluded that such a move would not be a useful or proportionate measure at this time. In our view, the wider international dimension is best served by the energetic and open exchange of information. We welcome the fact that information is already being exchanged at all levels, from the OECD to individual scientists, and urge the further facilitation of such exchanges.

4.71   Of the additional measures that we considered, we were most attracted by the development of some kind of *early warning system*, one that might be managed by the competent authorities for REACH or by a body or bodies authorised by them to do so. Indeed, as we confront the control dilemma, it seems to us that an early warning system incorporating reporting requirements is a vital component of governance.

4.72   We are aware that such reporting could be onerous, especially for small and medium-sized enterprises. Hence, we recommend that such reporting should be kept as simple as possible. We are attracted by the idea of a straightforward checklist aimed primarily at nanomaterials that are not currently captured by REACH. All importers or manufacturers of such materials, or products containing them (above some still-to-be-decided threshold for the quantities involved) that are not captured by REACH, would be required to complete the checklist in as much detail as they are able. It should be designed so as not to be onerous, should elaborate the special properties of the nanomaterials including the reason that they have been produced or incorporated in the product, and should also consider the pathways of environmental and human exposure throughout the entire life cycle of the product, not just at the point of use. In addition, the checklist might prompt explicit consideration of the consumer benefits of the product. We have received evidence that some leading manufacturers use similar checklists (developed as part of legal risk assessment requirements). **We recommend that the idea of a simple checklist as part of an early warning system be developed and defined further by the Government to investigate the potential for development amongst the wider materials community.**

4.73   We recognise that in making this recommendation, we are treating nanomaterials as a single group, apparently violating our own argument that it is their properties and functionalities that matter, not their size. But in the present state of ignorance, we see no other means of gathering and monitoring information that would allow society to move to regulations based on properties.

4.74   Experience suggests that, whilst we see a role for some kinds of voluntary initiatives as noted above (4.66), checklist reporting will have to be compulsory if it is to be effective. It is clear, for example, that the current voluntary reporting scheme for engineered nanoscale materials[39] operated by Defra has not worked. During its two years of operation, Defra has received only nine submissions.[40] **Hence, we recommend that Defra should make nanomaterials reporting mandatory.**

4.75   Companies, researchers and regulatory authorities all gather information about possible and emerging risks. The sharing of such information (including evidence that risks are minimal) must also be an important part of any early warning system. Whilst there are some genuine problems concerning commercial confidentiality, it must be possible to devise schemes in which information provided to the competent authorities is shared (anonymously if need be) with others who need to know.

4.76   Hence, **we recommend that the Government impose an additional legal duty on companies to report at the earliest opportunity to the competent authorities any reasonable suspicion that a material presents a risk to people or the environment. Compliance with this requirement should offer duty holders a degree of immunity from criminal liability, should problems associated with the nanomaterials arise in future.** A possible model is provided by Section 16 of the Health and Safety at Work Act 1974, which provides that the Health and Safety Commission can approve codes of practice. Failure to abide by such a code is not, as such, a criminal offence but it can be used as evidence that there is a breach of a regulation to which the code relates. Furthermore, guidance may be issued and, while compliance with this does not provide immunity from prosecution, the policy view is that an employer following the guidance will not commit a breach.

4.77   Although compliance with regulatory requirements does not provide immunity from civil law, it could in principle form part of a defence against a claim in negligence or nuisance. Reasonable foreseeability of harm is an important element in these torts and, as was held in Cambridge Water Company vs. Eastern Leather,[41] requires recognition of the toxic nature of the pollutant in question, the ability to detect its presence and an understanding of its likely pathways in the environment. If, following an honest disclosure under the checklist and risk reporting duties, the regulatory authority has taken no action, this would argue against the foreseeability of harm. The problem at present (as noted in Chapter 3) is that the means for detecting nanoparticles in the environment and attributing them to particular sources are underdeveloped or non-existent.

4.78   Pervading all of these activities, therefore, is a requirement for a robust programme of *environmental monitoring*, using new techniques to detect manufactured nanoparticles in living organisms and the environment (Chapter 3). Monitoring, as we have argued elsewhere,[42] is an essential component of any early warning system. While 'blanket' monitoring of the environment is not practicable, targeted monitoring is highly desirable. Obvious points for surveillance might include sewage outfalls, river water and sediments downstream from major conurbations, coastal marine sediments and sediment-feeding organisms. Detection of significant quantities of a nanomaterial in a top predator (pike or otters for example) could also be a cause for concern.

4.79 **We recommend that environmental monitoring to detect manufactured nanoparticles should be the responsibility of the Environment Agency in England and Wales, SEPA in Scotland and the Northern Ireland Environment Agency to ensure that robust processes are used.** While we recognise that some environmental media are monitored by other bodies (e.g. air quality monitoring by local authorities), we believe that the complexity and cost of the monitoring likely to be required necessitates a more focused approach and responsibility. This is not to neglect the importance of information sharing across responsible bodies.

4.80 There is a further difficulty in making causal connections between the presence of nanomaterials in the environment and any observed changes or deleterious effects, one consequence of which is that, in the present state of knowledge, any civil action would be unlikely to succeed. We were told that it was not possible to 'mark' nanomaterials so that they could be traced back to a particular manufacturer or importer.[43] If damage is caused therefore, any remediation and compensation would fall to the public purse.

4.81 In this section we have reviewed a number of measures that might improve the knowledge base and strengthen or supplement regulatory arrangements in Europe and the UK. We envisage that knowledge about nanomaterials, their behaviour in organisms and the environment, and their potential risks will accumulate, and that over time REACH and the sector-specific regulations will be adjusted (as discussed in the previous section, 4.35-4.52). In the meantime, there is clearly a need for vigilance. That is why we propose that attention should be given to the development of an early warning system as a supplementary measure to ensure as far as possible that significant and irreversible harm will not occur.

4.82 We recognise, however, that the measures advocated above present only a partial response to the challenges posed even by first and second generation nanomaterials, that nanomaterials themselves represent a particular example of the issues raised by the wider class of novel materials, and that both categories, as we noted at the outset of this chapter, raise and exemplify fundamental questions about the ways in which modern societies shape and respond to rapidly emerging technologies.

GOVERNING EMERGENT TECHNOLOGIES

4.83 There are no simple and straightforward solutions to the control dilemma. It is possible, and indeed essential, to narrow the gaps through concerted efforts in research and by tightening and extending existing regulations. But the governance of emerging technologies in the face of ubiquity, ignorance and uncertainty must amount to much more than this.

4.84 Nanomaterials exemplify the kind of challenge for which attention to closing gaps in knowledge and regulation is necessary but insufficient. Effective governance will mean looking beyond traditional regulation for other, more imaginative solutions, often involving a wider range of actors and institutions than has been customary in the past. The aim must be to create *adaptive management systems* that can respond quickly and effectively as new information becomes available. We have explored a number of options in the previous sections but have clearly not exhausted the possibilities.

4.85   Ultimately however, as we noted in Chapter 1, many of the questions raised by developments like those in the field of novel materials are trans-scientific in nature. They extend beyond the (important) issues of risk and risk management to questions about the direction, application and control of innovation.[44] Indeed, it has been argued that "public misgivings over the purposes and interests behind innovations are often misunderstood as if they are concerns about safety as defined by regulatory science and expertise".[45] The more substantive challenge, therefore, is to find the means through which civil society can engage with the social, political and ethical dimensions of science-based technologies, and democratise their 'licence to operate'. It has been characterised as a challenge of moving beyond the governance of risk to the governance of innovation.[46]

4.86   This is not an easy task. It will demand the engagement of a wide range of different perspectives and, quite possibly, the establishment of new institutions (an aspect to which we return below). There is growing recognition of these requirements, and we are aware of energetic activity particularly in the areas of opinion gathering and public and stakeholder engagement with nanotechnologies.

4.87   Efforts have been made, for example, to gauge social perceptions in opinion polls.[47] Surveys suggest that most Americans and Europeans are unfamiliar with nanotechnologies but, when prompted, anticipate that the benefits might outweigh the risks. The most frequently expressed concerns relate to loss of privacy and to the risks that might be posed by artificial self-replicating organisms (effects normally associated with later generations of nanotechnologies). Findings indicate that there is greater acceptance of nanomaterials in some applications, such as medicine and environmental protection, than in others such as food.[48] However, survey research methods have obvious limitations under conditions of novelty and emergence, and perspectives on nanotechnologies have also been explored in a variety of deliberative forums.

4.88   In response to the Royal Society and Royal Academy of Engineering report,[49] the UK Government established a programme of work on nanotechnologies including three projects funded through the *Sciencewise* public engagement programme: 'Small Talk'; 'Nanodialogues'; and the Nanotechology Engagement Group. A further seven projects, not directly funded by government, underlined a concern to open up discussion of nanotechnologies to new perspectives.[50]

4.89   An overview of projects involving public engagement in the UK and the United States, including some of the above, identified four common emerging themes: an expectation that nanotechnologies will deliver benefits; anxiety about the management of unforeseen risks; concern that innovation would not be directed towards appropriate social goals; and a desire for science and technology policy to be more open to public involvement.[51] In some cases, ignorance, followed by initial enthusiasm about potential benefits, turned to unease after participants learned more about the technologies concerned. In one project, focus group participants, having interrogated websites and discussed the issues with family and friends, expressed concerns about nanoparticles entering and harming the body, their fears being exacerbated by the notion of 'invisibility'. The potential toxicity of nanomaterials was seen as symptomatic of rapid technological development in the face of uncertainties and ignorance about risks.[52]

4.90   The intensity of the public engagement effort is in part a reflection of the speed and scale of innovation. But it is driven also by concerns about possible societal responses to particular technologies. In some cases, this concern is motivated by a desire to avoid friction: we heard frequently in Japan, for example, that public engagement was aimed at increasing the acceptability of nanotechnologies.[53] Such motivation, whilst not always explicit, has also been an important factor behind public engagement initiatives in the UK, although the Government has stated its aims more broadly, in terms of building a society "confident about the governance, regulation and use of science and technology".[54]

4.91   We urge that the emphasis be placed on these broader objectives. The full value of engagement and deliberation will not be realised if these activities are seen primarily as an exercise in securing acquiescence to new technologies. Rather, they should constitute an important component in a system of innovation governance.

4.92   We welcome initiatives that contribute to the 'social intelligence' function of governance in this context, including opportunities for deliberation among a wide range of different groups and members of the public. We endorse the views of the Nanotechnology Engagement Group (NEG) (4.88) that research and policy in this field should be informed by public insights and concerns, and that scientists should have opportunities to reflect upon the wider social implications of their work.[55] We note that such views have won support from, amongst others, the Royal Society and the former Office of Science and Technology.

4.93   Experience has grown of a variety of mechanisms to facilitate public engagement and deliberation, including citizens' juries, consensus conferences and facilitated discussions. Examples of all of these in the context of nanotechnologies were included among the exercises mentioned above. In relation to research cultures, we note with interest the support given by the Societal Issues Panel of the Engineering and Physical Sciences Research Council (EPSRC) to building social priorities and concerns into strategic funding decisions. For example, it has supported a nanotechnology dialogue in 2008 to inform the development of a new research programme in this field.[56]

4.94   However, we acknowledge that the task is made particularly challenging in the case of nanomaterials, as with many emergent technologies, by the global innovation system involved. We are also well aware that any forum that is time limited and focused on single technologies or types of technology is subject to limitations of specific timing, representativeness and agenda framing, amongst others.

4.95   **Hence, we recommend that it is desirable to move beyond one-off public engagement 'projects' to recognise the importance of continual 'social intelligence' gathering and the provision of ongoing opportunities for public and expert reflection and debate. We see these functions as crucial if, as a society, we are to proceed to develop new technologies in the face of many unknowns.**

4.96 There is a growing, formal literature on how to achieve these ends, for example, through techniques such as 'Real-Time Technology Assessment' (RTTA)[57] and 'Constructive Technology Assessment' (CTA).[58] CTA developed in the Netherlands and is usually described as having three elements: socio-technical mapping combining stakeholder analysis and the systematic plotting of recent technical dynamics; early but controlled experimentation, through which unanticipated impacts can be identified and, if necessary, mitigated; and dialogue between innovators and the public, to articulate the demand side of technology development and innovation.

4.97 RTTA has a similar objective, but it is usually less focused on experimentation and more on the knowledge generation process itself. It makes use of more reflexive measures such as focus groups and scenario development to elicit values from multiple stakeholders and to explore alternative potential outcomes. It uses forms of social survey research to investigate how knowledge, perceptions and values are evolving over time to enhance communication and identify emerging problems. It integrates socio-technical mapping and dialogue with retrospective as well as prospective (scenario) analysis.

4.98 In the face of ignorance, uncertainty and a rapidly changing knowledge base, an important feature of systems supporting adaptive management like CTA and RTTA is scenario building – asking the 'what if?' questions and doing so through involvement of a variety of expert, public and local knowledges. This helps to ensure that the process is informed by a range of different perspectives and by social and ethical as well as scientific and technical considerations.

4.99 Another related possibility that has been put to us is that decisions about the development of new technologies would be strengthened considerably if they were subject to explicit reflection from a purposely designed, deliberative public forum, the results of which could directly inform government thinking and policy.[59] We have considerable sympathy for this view, although only if designed as an ongoing rather than a time-limited activity. One possibility is that of a standing deliberative forum, designed to inform policy on nanotechnology development, regulation and research.[60] It might be possible, in this context, to build upon Defra's Nanotechnologies Stakeholder Forum, which has good links with industry, civil society and academia. Indeed, we understand that this body is being merged with the Chemical Stakeholders' Forum, which in turn will be informed by a restructured Advisory Committee on Hazardous Substances. We welcome this move.

4.100 However, a deliberative forum of the kind envisaged would cover a broad spectrum of world views, and would have to be capable of exploring normative questions about the purpose, direction and control of innovation, as well as issues of risk and regulation. It would be a fundamental exploration of values and visions of the future of technologies, informed by the best science available at the time.[61] This kind of forum is different in concept to a stakeholder body, which assembles and seeks consensus among different interests.

4.101 There remains in our view a real question about whether the capacity for deliberation and (perhaps even more so) for public engagement in modern democracies is sufficient to sustain an approach that seeks to interrogate scientific and technological developments. The more specific the focus, the more numerous the cases will be, but as the focus becomes more generalised (looking at 'nanotechnologies' as a whole for example), the range of possible applications and implications threatens to make dialogue unmanageable. Informed and inclusive deliberation (however conducted) on a huge range of potential developments seems as distant a prospect as the resolution of many of the technical uncertainties identified elsewhere in this report.

4.102 A different approach to the governance of innovation – given the problems of deliberating emergent developments on a case-by-case basis – might be to begin with questions of principle, instead of working from technologies through to implications. A key task would be to consider which kinds of interventions in the human and non-human worlds, controlled by whom, might be deemed acceptable or problematic. Such principles could then act as a filter, directing attention to aspects of particular science-based innovations that seemed worthy of special scrutiny. Deliberation on these issues might be a role for a commission on emerging technologies and society, of the kind that has recently been proposed by Demos.[62] But it could also be undertaken by a body such as the deliberative forum considered above, if it concerned itself with a wide range of technological innovations.

4.103 Whatever institutional arrangements are adopted in pursuit of plurality and social intelligence, we are convinced that more rigorous attention needs to be paid to the treatment of the outputs. It seems to us that enthusiasm to be seen to engage has sometimes run ahead of any real commitment or institutional capacity not only to support the activities adequately but most importantly to make intelligent and transparent use of the findings, especially if the latter raise fundamental questions about the direction and development of innovation. Genuine 'upstream engagement', the outputs of which influence science and technology policy at an early stage,[63] has proved elusive, and is particularly challenging under conditions of ubiquity when world views vary widely across countries and cultures.[64]

4.104 We return, finally, to the control dilemma. We have argued that this dilemma clearly confronts us in the case of nanomaterials, our focus in this report, and that our response should be to strive towards an open and adaptive system of governance grounded in reflective and informed technical and social intelligence. Such a regime, while encouraging appropriate innovation, would seek to avoid technological inflexibility, would be vigilant and would be capable of intervening selectively but decisively when developments threatened humans or the non-human environment.

4.105 We have argued that adaptive innovation governance for nanomaterials would be served by modifying and extending the existing regulatory framework as a matter of urgency, and by developing an early warning system which must include robust arrangements for monitoring. But, as in other fields characterised by ignorance, uncertainty and ubiquity, regulatory measures will not resolve the dilemma, and therefore they must be complemented (and informed) by the full range of perspectives on innovation. It is to these ends that we have made our recommendations, some of which we consider applicable beyond nanomaterials to novel materials in general, and indeed to the governance of wider categories of emergent technologies.

Chapter 5

SUMMARY OF RECOMMENDATIONS

5.1 In this study we looked at issues related to innovation in the materials sector and in particular the challenges and benefits arising from the introduction of novel materials. This was prompted by concerns about potential releases to the environment from industrial applications of metals and minerals that have not previously been widely used and from the use of manufactured nanomaterials in a wide variety of products and applications. In the event, the evidence we received was almost entirely focused on the latter and we therefore decided to use nanomaterials as an exemplar to investigate the current governance frameworks surrounding innovation and protection of human health and the environment in this sector.

5.2 In Chapter 2 of the report we looked at the properties of nanomaterials currently being used or developed and the functionalities derived from those properties which allow the introduction of new products or improved products and performance. We also looked at the potential pathways by which these materials could enter the environment and present potential hazards to the environment and to people throughout their life cycle.

5.3 Chapter 3 dealt with the potential health and environmental impacts which flow from the properties of nanomaterials and we concluded that there is a plausible case for concern about some (but not all) classes of nanomaterials. Examples of potentially harmful nanomaterials include nanosilver, carbon nanotubes and Buckminsterfullerenes (C_{60}). However, we are very conscious of the extent to which knowledge about the potential health and environmental impacts of nanomaterials lags significantly behind the pace of innovation and these areas of concern could change as new scientific information arises. This is an area of considerable uncertainty.

5.4 In Chapter 4 we looked at the current governance arrangements and the ways in which these deal with both uncertainty and ignorance about possible risks to the environment, organisms and human health. We focused on fairly narrow issues in relation to governance in the materials and chemicals sectors and the particular challenges to governance posed by nanomaterials. Some of our recommendations are specifically directed at issues which relate to manufactured nanomaterials. In particular, that the nanoform of a chemical may have significantly different properties to its bulk form. In the longer term, we are also concerned that more sophisticated third and fourth generation nanoproducts may represent a further step change in functionalities and properties, which would be even more difficult to capture in a regulatory system primarily focused on the bulk chemical properties of the material. We concluded that new governance arrangements are necessary to deal with ignorance and uncertainty in this rapidly developing area. It has not escaped our notice that, in principle, these arrangements could also apply in areas of technology other than novel materials where similar issues might arise.

5.5   Our recommendations reflect three main priorities, namely:

- Functionality: we need to focus on the properties and functionalities of specific nanomaterials as the key driver rather than treat all materials in the size range as one single class.

- Information: we need to establish directed research programme on the properties and functionalities of materials in order to inform risk assessment and risk management strategies.

- Adaptive management: we need to recognise the degree of ignorance and uncertainty and the time it will take to address these (insofar as they can be addressed). We also need to develop flexible and resilient forms of adaptive management to allow us to handle such difficult situations and emergent technologies.

These issues underlie the specific recommendations in Chapters 3 and 4, which are listed below.

ENVIRONMENTAL AND HEALTH IMPACTS

5.6   The research requirements highlighted in Chapter 3 of this report need to be undertaken on a more systematic and strategic basis, which is difficult to deliver under response mode funding as currently used in the UK as the main driver. We appreciate that the Government did not wish to take up the recommendations of the Royal Society and Royal Academy of Engineering report for a new research centre, but **we strongly recommend a more directed, more co-ordinated and larger response led by the Research Councils to address the critical research needs raised by this report, with emphasis on regulatory and policy programmes (3.114).**

5.7   These needs include:

- The validation of *in vitro* tests against *in vivo* models.

- Evaluation of methodologies for predicting the likely fate and effects of nanomaterials based on their physical and chemical properties as well as their novel properties and, where possible, the development of exposure scenarios.

- Based on the significant gap in our knowledge, the programme of directed research should ensure a concerted and co-ordinated effort is made to better understand the principles that drive the toxicity of manufactured nanomaterials and how individual properties interact to enhance or diminish toxicity profiles both *in vitro* and *in vivo* with a long-term objective of developing predictive toxicology.

- The enhancement of *in situ* monitoring and surveillance methods to provide early warnings of unexpected effects of novel materials and to permit timely remedial action.

- The research programme should pave the way for much greater interdisciplinary co-operation, including co-operation between those engaged in medical toxicology and those in ecotoxicology, so as to enhance the development of robust test systems and also to act as a catalyst for early warnings from observations on lower organisms to be extrapolated to humans (3.115).

5.8   We recommend that urgent attention is given to undergraduate and postgraduate training in toxicology across all of its domains and that the Department for Innovation, Universities and Skills (DIUS), the university sector and the professional societies that represent medical toxicologists and ecotoxicologists establish new initiatives to build multidisciplinary capacity in this field (3.117).

GOVERNANCE

5.9   We recommend that in any revisions to existing regulations, the relevant authorities should focus specifically on the properties and functionalities of nanomaterials, rather than size. Since these properties and functionalities will often differ substantially from those of the bulk material, strict chemical equivalence does not preclude the need for a separate risk assessment (4.45).

5.10 We recommend that the UK Government should press the European Commission to proceed with urgency, in consultation with Member States, the European Chemicals Agency and the Scientific Committee on Emerging and Newly Identified Health Risks (SCENIHR), to review REACH and the product- or sector-specific regulations. The object of the review should be to amend the regulations to facilitate their effective application to nanomaterials and the provision of adequate testing arrangements (4.47).

5.11 We recommend the establishment of clear priorities for testing, beginning with those nanoparticles with functionality which suggests that they might pose the greatest risk of harm to the environment or to human health (4.48).

5.12 We recommend that, as REACH is adapted to meet the challenges presented by nanomaterials, particular attention should be given to the issue of weight thresholds. In view of the persistent uncertainties involved, a precautionary approach should be adopted when determining new, lower thresholds for nanomaterials (4.51).

5.13 We recommend that responsible organisations set up structured systems to keep a watching brief on the development of novel materials and to enhance the sharing of information and the opportunities to work together to identify and manage emerging problems (4.64).

5.14 We recommend that the idea of a simple checklist as part of an early warning system be developed and defined further by the Government to investigate the potential for development amongst the wider materials community (4.72).

5.15 We recommend that the Department for Environment, Food and Rural Affairs (Defra) should make nanomaterials reporting mandatory (4.74).

5.16 We recommend that the Government impose an additional legal duty on companies to report at the earliest opportunity to the competent authorities any reasonable suspicion that a material presents a risk to people or the environment. Compliance with this requirement should offer duty holders a degree of immunity from criminal liability, should problems associated with the nanomaterials arise in future (4.76).

5.17 We recommend that environmental monitoring to detect manufactured nanoparticles should be the responsibility of the Environment Agency in England and Wales, the Scottish Environment Protection Agency (SEPA) and the Northern Ireland Environment Agency to ensure that robust processes are used (4.79).

5.18 We recommend that it is desirable to move beyond one-off public engagement 'projects' to recognise the importance of continual 'social intelligence' gathering and the provision of ongoing opportunities for public and expert reflection and debate. We see these functions as crucial if, as a society, we are to proceed to develop new technologies in the face of many unknowns (4.95).

ALL OF WHICH WE HUMBLY SUBMIT FOR

YOUR MAJESTY'S GRACIOUS CONSIDERATION

John Lawton *Chairman*

Nicholas Cumpsty

Michael Depledge

Paul Ekins

Ian Graham-Bryce

Stephen Holgate

Jeffrey Jowell

Peter Liss

Susan Owens

Judith Petts

Steve Rayner

John Speirs

Janet Sprent

Lynda Warren

Tom Eddy   *Secretary*

Jo Bray

Jon Freeman

Noel Nelson   *Secretariat*

Kay Sadanand

Laura Pleasants

80

References

Chapter 1

1   Royal Commission on Environmental Pollution (2003). *Twenty-fourth Report: Chemicals in Products: Safeguarding the Environment and Human Health*. TSO, London. Available at: http://www.rcep.org.uk

2   Evidence from Professor Ken Donaldson, University of Edinburgh, July 2007.

3   Evidence from Dr Andrew Maynard, Woodrow Wilson Center, July 2007.

4   European Environment Agency (EEA) (2001). *Late Lessons from Early Warnings: The precautionary principle 1896-2000*. Environmental Issue Report No. 22. EEA, Copenhagen.

5   Poland, C.A., Duffin, R., Kinloch, I., Maynard, A., Wallace, W.A.H., Seaton, A., Stone, V., Brown, S., MacNee, W. and Donaldson, K. (2008). Carbon nanotubes introduced into the abdominal cavity of mice show asbestos-like pathogenicity in a pilot study. *Nature Nanotechnology*, **3**, 423-428.

6   Environment Agency Interim Advice on wastes containing unbound carbon nanotubes. Issued 19 May 2008. Available at: http://www.nerc.ac.uk/research/programmes/nanoscience/events/documents/nano-waste.pdf. Accessed 15 July 2008.

7   Weinberg, A. (1972). Science and trans-science. *Minerva*, **10**(2), 209-222.

8   Roco, M. and Bainbridge, W. (Eds.) (2001). *Societal Implications of Nanoscience and Nanotechnology*. National Science Foundation, Arlington, Virginia;
US National Research Council (2002). *Small Wonders, Endless Frontiers*. National Academies Press, Washington DC;
Nordmann, A. (Rapporteur) (2004). *Converging Technologies – Shaping the Future of European Societies*. A report by the High Level Expert Group (HLEG) on Foresighting the New Technology Wave. European Commission, Brussels;
Royal Society and Royal Academy of Engineering (RS/RAE) (2004). *Nanoscience and Nanotechnologies: Opportunities and uncertainties*. Royal Society Policy Document 19/04.

9   Collingridge, D. (1980). *The Social Control of Technology*. Francis Pinter, New York.

10   RS/RAE (2004), page 84.

Chapter 2

1   Information on the Lycurgus cup at the British Museum available at: http://www.britishmuseum.org/explore/highlights/highlight_objects/pe_mla/t/the_lycurgus_cup.aspx. Accessed 15 July 2008.

2   Nel, A., Xia, T., Madler, L. and Li, N. (2006). Toxic potential of materials at the nanolevel. *Science*, **311**(5761), 622-627.

3   Personal communication with the US Food and Drug Administration, January 2008.

4   Photograph © Dr Andrei Khlobystov, University of Nottingham.

5   Royal Society and Royal Academy of Engineering (RS/RAE) (2004). *Nanoscience and Nanotechnologies: Opportunities and uncertainties*. Royal Society Policy Document 19/04.

6   Diagram of a fullerene from Peter Unwin, Department of Chemistry, Imperial College of Science, Technology and Medicine. Available at: http://www.ch.ic.ac.uk/local/projects/unwin/c84.GIF. Accessed 28 July 2008.

7   Hochella, M.F., Lower, S.K., Maurice, P.A., Penn, R.L., Sahai, N., Sparks, D.L. and Twining, B.S. (2008). Nanominerals, mineral nanoparticles and earth systems. *Science*, **319** (5870), 1631-1635.

8   Personal communication with Dr Andrew Maynard, Woodrow Wilson Center, June 2008.

9   Feynman, R.P. (1959). *There's Plenty of Room at the Bottom*. American Physical Society, California Institute of Technology. Available at: http://www.zynex.com/nanotech/feynman.html. Accessed 15 July 2008.

10   Li, L., Fan, M., Brown, R., van Leeuwen, J., Wang, J., Wang, W., Song, Y. and Zhang, P. (2006). Synthesis, properties, and environmental applications of nanoscale iron-based materials: A review. *Critical Reviews in Environmental Science and Technology*, **36**(5), 405-431.

11 Doty, R.C., Tshikhudo, T.R., Brust, M. and Fernig, D.G. (2006). Extremely stable water-soluble Ag nanoparticles. *Chemistry of Materials*, **17**, 4630-4635.

12 Hoppe, C.E., Lazzari, M., Pardinas-Blanco, I. and Lopex-Quintela, M.A. (2006). One-step synthesis of gold and silver hydrosols using poly(N-vinyl-2-pyrrolidone) as a reducing agent. *Langmuir*, **22**, 7027-7034.

13 Hester, R.E. and Harrison, R.M. (2007). *Nanotechnology: Consequences for Human Health and the Environment.* Issues in Environmental Science and Technology, Volume 24. Royal Society of Chemistry Publishing.

14 Hunter, K.A. and Liss, P.S. (1979). The surface charge of suspended particles in estuarine and coastal waters. *Nature*, **282**, 823-835.

15 Nightingale, P., Smith, A., van Zwanenberg, P., Rafols, I. and Morgan, M. (2008). *Nanomaterial Innovations Systems: Their Structure, Dynamics and Regulation.* Science and Technology Policy Research Unit (SPRU), University of Sussex. A report for the Royal Commission on Environmental Pollution. Available at: http://www.rcep.org.uk

16 Renn, O. and Roco, M. (2006). *White Paper on Nanotechnology Risk Governance.* International Risk Governance Council (IRGC).

17 Personal communication with Professor George Smith, University of Oxford, July 2006.

18 Nightingale *et al.* (2008).

19 Pierstorff, E. and Ho, D. (2007). Monitoring, diagnostic, and therapeutic technologies for nanoscale medicine. *J. Nanosci. Nanotechnol.*, **7**(9), 2949-2968.

20 Streicher, R.M., Schmidt, M. and Fiorito, S. (2007). Nanosurfaces and nanostructures for artificial orthopedic implants. *Nanomed.*, **2**(6), 861-874.

21 Patolsky, F., Zheng, G. and Lieber, C.M. (2006). Nanowire sensors for medicine and the life sciences. *Nanomed.*, **1**(1), 51-65.

22 *Ibid.*

23 Béduneau, A., Saulnier, P. and Benoit, J.P. (2007). Active targeting of brain tumors using nanocarriers. *Biomaterials*, **28**(33), 4947-4967.

24 Najlah, M. and D'Emanuele, A. (2007). Synthesis of dendrimers and drug-dendrimer conjugates for drug delivery. *Curr. Opin. Drug Discov. Devel.*, **10**(6), 756-767.

25 Sumer, B. and Gao, J. (2008). Theranostic nanomedicine for cancer. *Nanomed.*, **3**(2), 137-140**.**

26 Igarashi, E. (2008). Factors affecting toxicity and efficacy of polymeric nanomedicines. *Toxicol. Appl. Pharmacol.*, **229**(1), 121-134.

27 *Woodrow Wilson Center Project on Emerging Nanotechnologies Consumer Products Inventory.* Available at: http://www.nanotechproject.org/inventories/consumer/. Accessed 5 July 2008.

28 Renn and Roco (2006).

29 *Woodrow Wilson Center Project on Emerging Nanotechnologies Consumer Products Inventory.* Available at: http://www.nanotechproject.org/inventories/consumer/. Accessed 5 July 2008.

30 Nightingale *et al.* (2008).

31 *Ibid.*

32 *Ibid.*

33 *Ibid.*

34 Oakdene Hollins (2007). *Environmentally Beneficial Nanotechnologies.* A report commissioned by the Department for Environment, Food and Rural Affairs (Defra). Available at: http://defraweb/environment/nanotech/policy/index.htm. Accessed 15 July 2008.

35 US National Science and Technology Council (2000). *National Nanotechnology Initiative: The initiative and its implementatio*n. Washington DC. Available at: http://www.nano.gov/html/res/nni2.pdf. Accessed 28 July 2008.

36 Renn and Roco (2006).

37 Nightingale *et al.* (2008).

38 *Ibid.*

39 *Ibid.*

40   Biswas, P. and Wu, C.-Y. (2005). Nanoparticles and the environment. *Journal of the Air and Waste Management Association*, **55**(6), 708-746;

US Environmental Protection Agency (2005). *Nanotechnology White Paper*. Available at: http://www.epa.gov/OSA/pdfs/EPA_nanotechnology_white_paper_external_review_draft_12-02-2005.pdf. Accessed 23 July 2008;

Boxall, A.B.A., Tiede, K. and Chaudry, Q. (2007). Engineered nanomaterials in soils and water: How do they behave and could they pose a risk to human health? *Nanomedicine*, **2**, 919-927.

41   Powers, K., Brown, S., Krishna, V., Wasdo, V., Wasdo, S., Moudgil, B. and Roberts, S. (2006). Characterisation of nanoscale particles for toxicological evaluation. *Toxicol. Sci.*, **90**, 296-303.

42   Adam, L., Lyon, D. and Alvarez, P. (2006). Comparative ecotoxicity of nanoscale TiO_2, SiO_2 and ZnO_2 water suspensions. *Water Res.*, **40**, 3527-3532.

43   Burleson, D.J., Driessen, M.D. and Penn, R.L. (2004). On the characterisation of environmental nanoparticles. *J. Environ. Sci. Health Part A*, **39**, 2707-2753.

44   Personal communication with Mark Chappell, US Army Corps of Engineers, and Chris Metcalfe, Trent University, Peterborough, Canada, April 2008.

45   Hansen, S.F., Larsen, B.H., Olsen, S.I. and Baun, A. (2007). Categorisation framework to aid hazard identification of nanomaterials. *Nanotoxicology*, **1**(3), 243-250.

46   Brant, J., Lecoanet, H. and Wiesner, M.R. (2005). Aggregation and deposition characteristics of fullerene nanoparticles in aqueous systems. *Journal of Nanoparticle Research*, **7**(4-5), 545-553;

Dunphy-Guzman, K.A., Taylor, M.R. and Banfeils, L.F. (2006). Environmental risks of nanotechnology: National Nanotechnology Initiative funding 2000-2004. *Environ. Sci. Technol.*, **40**, 1401-1407;

Hyung, H., Fortner, J.D., Hughes, J.B. and Kim, J.H. (2007). Natural organic matter stabilizes carbon nanotubes in the aqueous phase. *Environ. Sci. Technol.*, **41**(1), 179-184;

personal communication with Mark Chappell, US Army Corps of Engineers, and Chris Metcalfe, Trent University, Peterborough, Canada, April 2008.

47   Lead, J.R. and Wilkinson, K.J. (2006). Aquatic colloids and nanoparticles: Current knowledge and future trends. *Environmental Chemistry*, **3**, 159-171.

48   Fortner, J., Lyon, D. and Sayes, C. (2005). C-60 in water: Nanocrystal formation and microbial response. *Environ. Sci. Technol.*, **39**, 4307-4316;

Phenrat, T., Saleh, N., Sirk, K., Tilton, R. and Lowrey, G. (2007). Aggregation and sedimentation of aqueous nanoscale zerovalent iron dispersions. *Environ. Sci. Technol.*, **41**, 284-290;

Lead, J.R., Smith, E.L., Scott-Fordsmand, J.J., Baun, A., Handy, R.D., Slaveykova, V.I., Tyler, C.R., von der Kammer, F., Benedetti, M., Boxall, A., Brust, M., Cumpson, P., Fernandes, T., Hassellov, M., Henry, T.B., Holbrook, R.D., Kookana, R., Masion, A., McClellan-Green, P., Nelson, N., Owen, R., Park, B., Garrod, J., Valsami-Jones, E. and Vincent, B. (2008). *Linking the physico-chemical characteristics and ecotoxicology of manufactured nanomaterials in aquatic and terrestrial environments*. Paper in preparation for publication.

49   Stolpe, B. and Hassellov, M. (2007). Changes in size distribution of fresh water nanoscale colloidal matter and associated elements on mixing in seawater. *Geochim. Cosmochim. Acta*, **71**, 3292-3301.

50   Handy, R.D. and Shaw, B.J. (2007). Ecotoxicity of nanomaterials to fish: Challenges for ecotoxicity testing. *Integrated Environmental Assessment and Management*, **3**(3), 458-460.

51   Stolpe and Hassellov (2007).

52   Handy, R. and Owen, R. (Guest Editors) (2008). Special Issue on Ecotoxicology, Chemistry and Risk Assessment of Nanoparticles. *Ecotoxicology*, Issue Number 5.

53   Personal communication with Mark Chappell, US Army Corps of Engineers, and Chris Metcalfe, Trent University, Peterborough, Canada, April 2008.

54   Tungittiplakorn, W., Lion, L.W., Cohen, C. and Kim, J.Y. (2004). Engineered polymeric nanoparticles for soil remediation. *Environ. Sci. Technol.*, **38**(5), 1605-1610.

55   Lecoanet, H.F., Bottero, J.Y. and Wiesner, M.R. (2004). Laboratory assessment of the mobility of nanomaterials in porous media. *Environ. Sci. Technol.*, **41**, 4465-4470.

56   Galloway, T. (2008). *Study of Novel Materials: Toxicology Literature Review.* Consultancy study in support of the Royal Commission on Environmental Pollution's Twenty-seventh Report on Novel Materials. School of Biosciences, Hatherly Laboratories, University of Exeter. Available at: http://www.rcep.org.uk

57   Lead *et al.* (2008).

58   Moore, M.N. (2006). Do nanoparticles present ecotoxicological risks for the health of the aquatic environment? *Environ. International*, **32**, 967-976.

59   Kamat, J.P., Devasagayam, T., Priyadarsini, K., Mohan, H. and Mittal, J.P. (1998). Oxidative damage induced by the fullerene C-60 on photosensitisation in rat liver microsomes. *Chem.-Biol. Interact.*, **114**, 145-159.

60   Derfus, A.M., Chan, W. and Bhatia, S. (2004). Probing the cytotoxicity of semiconductor quantum dots. *Nano. Lett.*, **4**, 11-18.

61   Owen, R. and Depledge, M.H. (2005). Nanotechnology and the environment: Risks and rewards. *Marine Pollution Bulletin*, **50**(6), 609-612.

62   Diagram of a typical life cycle adapted from http://www.ami.ac.uk/courses/topics/0109_lct/. Accessed 17 July 2008.

Chapter 3

1   European Commission Scientific Committee on Emerging and Newly Identified Health Risks (SCENIHR) (2007). *Opinion on the appropriateness of the risk assessment methodology in accordance with the technical guidance documents for new and existing substances for assessing the risks of nanomaterials.* Available at: http://ec.europa.eu/health/ph_risk/committees/04_scenihr/docs/scenihr_o_010.pdf. Accessed 31 July 2008.

2   Linkov, I. and Satterstrom, F.K. (2008). Nanomaterial risk assessment and risk management: Review of regulatory frameworks. In: *Real-Time and Deliberative Decision Making.* Editors: I. Linkov, E. Ferguson and V.S. Magar. Springer, the Netherlands. Reproduced by kind permission of the lead author.

3   Royal Society and Royal Academy of Engineering (RS/RAE) (2004). *Nanoscience and Nanotechnologies: Opportunities and uncertainties.* Royal Society Policy Document 19/04.

4   *Ibid.*

5   Holbrook, D.R., Murphy, K.E., Morrow, J.B. and Cole, K.D. (2008). Trophic transfer of nanoparticles in a simplified invertebrate food web. *Nature Nanotechnology*, **3**, 352-355.

6   Personal communication with Dr Jamie Lead, University of Birmingham, April 2008.

7   Helland, A., Wick, P., Koehler, A., Schmid, K. and Som, C. (2007). Reviewing the environmental and human health knowledge base of carbon nanotubes. *Environ. Health Perspect.*, **115**(8), 1125-1131.

8   Thompson, R.C., Olsen, Y., Mitchell, R.P., Davis, A., Rowland, S.J., John, A.W.G., McGonigle, D. and Russell, A.E. (2004). Lost at sea: Where does all the plastic go? *Science*, **304**, 838.

9   Henn, K.W. and Waddill, D.W. (2006). Utilization of nanoscale zero-valent iron for source remediation – a case study. *Remediation*, **16**(2), 57-77.

10   Vo-Dinh, T. (2004). Biosensors, nanosensors and biochips: Frontiers in environmental and medical diagnostics. *The 1st International Symposium on Micro and Nano Technology*, 14-17 March 2004, Honolulu, Hawaii, USA. Available at: http://www.ornl.gov/~webworks/cppr/y2001/pres/119772.pdf. Accessed 31 July 2008.

11   Personal communication with Dr Jeff Steevens, US Army Corps of Engineers, April 2008.

12   Bontidean, I. (1998). Detection of heavy metal ions at femtomolar levels using protein-based biosensors. *Anal. Chem.*, **70**, 4162-4169.

13   Feng, J., Shan, G., Maquieira, S., Koivunen, M.E., Bing, G., Hammock, B.B. and Kennedy, I.M. (2003). Functionalised europium oxide nanoparticles used as a fluorescent label in an immunoassay for atrazine. *Anal. Chem.*, **75**, 5282-5286.

14 United Nations Environment Programme (UNEP) (2007). *Emerging Challenges: Nanotechnology and the Environment. UNEP GEO Year Book 2007.* Available at: www.unep.org/geo/yearbook/yb2007/. Accessed 28 July 2008.

15 Service, R.F. (2008). Don't sweat the small stuff. *Science*, **320**, 1584-1585.

16 European Environment Agency (EEA) (2001). *Late Lessons from Early Warnings: The precautionary principle 1896-2000.* Environmental Issue Report No. 22. EEA, Copenhagen.

17 *Woodrow Wilson Center Project on Emerging Nanotechnologies Consumer Products Inventory.* Available at: http://www.nanotechproject.org/inventories/consumer/. Accessed 5 July 2008.

18 Evidence from Professor Michael N. Moore, Plymouth Marine Laboratory, July 2007.

19 Depledge, M.H. (1993). Ecotoxicology: A science or a management tool? *Ambio (Royal Swedish Academy of Sciences)*, **22**, 51-52.

20 Holbrook *et al.* (2008).

21 Oberdörster, E. (2004). Manufactured nanomaterials (fullerenes, C_{60}) induce oxidative stress in the brain of juvenile largemouth bass. *Environmental Health Perspectives*, **112**(10), 1058-1062.

22 Brant, J., Leocanet, H. and Wiesner, M.R. (2005). Aggregation and deposition characteristics of fullerene nanoparticles in aqueous systems. *Journal of Nanoparticle Research*, **7**(4-5), 545-553.

23 Henry, T.B., Menn, F., Fleming, J.T., Wilgus, J., Compton, R.N. and Sayler, G.S. (2007). Attributing effects of aqueous C_{60} nano-aggregates to tetrahydrofuran decomposition products in larval zebrafish by assessment of gene expression. *Environmental Health Perspectives*, **115**(7), 1059-1065.

24 Lead, J.R., Smith, E.L., Scott-Fordsmand, J.J., Baun, A., Handy, R.D., Slaveykova, V.I., Tyler, C.R., von der Kammer, F., Benedetti, M., Boxall, A., Brust, M., Cumpson, P., Fernandes, T., Hassellov, M., Henry, T.B., Holbrook, R.D., Kookana, R., Masion, A., McClellan-Green, P., Nelson, N., Owen, R., Park, B., Garrod, J., Valsami-Jones, E. and Vincent, B. (2008). *Linking the physico-chemical characteristics and ecotoxicology of manufactured nanomaterials in aquatic and terrestrial environments.* Paper in preparation for publication.

25 Boxall, A., Chaudhry, Q., Sinclair, C., Jones, A., Aitken, R., Jefferson, B. and Watts, C. (2007a). *Current and Future Predicted Environmental Exposure to Engineered Nanoparticles.* Report by the Central Science Laboratory for the Department for Environment, Food and Rural Affairs (Defra).

26 Handy, R. and Owen, R. (Guest Editors) (2008). Special Issue on Ecotoxicology, Chemistry and Risk Assessment of Nanoparticles. *Ecotoxicology*, Issue Number 5.

27 Lewinski, N., Colvin, V. and Drezek, R. (2008). Cytotoxicity of nanoparticles. *Small*, **4**(1), 26-49.

28 Personal communication with Mark Chappell, US Army Corps of Engineers, and Chris Metcalfe, Trent University, Peterborough, Canada, April 2008.

29 Zhang, X., Sun, H., Zhang, Z., Niu, Q., Chen, Y. and Crittenden, J.C. (2007). Enhanced bioaccumulation of cadmium in carp in the presence of titanium dioxide nanoparticles. *Chemosphere*, **67**, 160-166.

30 Baun, A., Sorensen, S.N., Rasmussen, R.F., Hartmann, N.B. and Koch, C.B. (2008). Toxicity and bioaccumulation of xenobiotic organic compounds in the presence of aqueous suspensions of aggregates of nano-C(60). *Aquat. Toxicol.*, **86**(3), 379-387. E-pub. 2007.

31 Lead, J.R. and Wilkinson, K.J. (2006). Aquatic colloids and nanoparticles: Current knowledge and future trends. *Environmental Chemistry*, **3**, 159-171.

32 Lead *et al.* (2008).

33 Lam, C.W., James, J.T., McCluskey, R. and Hunter, R.L. (2004). Pulmonary toxicity of single-wall carbon nanotubes in mice 7 and 90 days after intratracheal instillation. *Toxicol. Sci.*, **77**(1), 126-134.

34 Wu, Y., Hudson, J., Lu, Q., Moore, J., Mount, A., Rao, A.M., Alexov, E. and Ke, P.C. (2006). Coating single-walled carbon nanotubes with phospholipids. *J. Phys. Chem. B*, **110**(6), 2475-2478.

35 Helland *et al.* (2007).

36 Handy and Owen (2008).

37   Galloway, T. (2008). *Study of Novel Materials: Toxicology Literature Review*. Consultancy study in support of the Royal Commission on Environmental Pollution's Twenty-seventh Report on Novel Materials. School of Biosciences, Hatherly Laboratories, University of Exeter. Available at: http://www.rcep.org.uk

38   Kloepfer, J.A., Mielke, R.E. and Nadeau, J.L. (2005). Uptake of CdSe/ZnS quantum dots into bacteria via purine-dependent mechanisms. *Appl. Env. Microb.*, **71**, 2548-2557.

39   *Ibid.*

40   Personal communication with Mark Chappell, US Army Corps of Engineers, and Chris Metcalfe, Trent University, Peterborough, Canada, April 2008.

41   Lyon, D.Y., Adams, L.K., Falkner, J.C. and Alvarez, P.J. (2006). Antibacterial activity of fullerene water suspensions: Effects of preparation method and particle size. *Environ. Sci. Technol.*, **40**(14), 4360-4366.

42   Tong, Z., Bischoff, M., Nies, L., Applegate, B. and Turco, R.F. (2007). Impact of fullerene (C_{60}) on a soil microbial community. *Environ. Sci. Technol.*, **41**, 2985-2991.

43   Nyberg, L., Turco, R.F. and Nies, L. (2008). Assessing the impact of nanomaterials on anaerobic microbial communities. *Environ. Sci. Technol.*, **42**, 1938-1943.

44   Navarro, E., Baun, A., Behra, R., Hartmann, N.B., Filser, J., Miao, A.J., Quigg, A., Santschi, P.H. and Sigg, L. (2008). Environmental behavior and ecotoxicity of engineered nanoparticles to algae, plants, and fungi. *Ecotoxicology*, **17**(5), 372-386.

45   Formina, M., Charnock, J.M., Hillier, S., Alvarez, R., Livens, F. and Gadd, G.M. (2008). Role of fungi in the biogeochemical fate of depleted uranium. *Curr. Biol.*, **18**(9), R375-377.

46   Scarano, G. and Morelli, E. (2003). Properties of phytochelatin-coated CdS nanocrystallites formed in a marine phytoplanktonic alga (*Phaeodactylum tricornutum*, Bohlin) in response to Cd. *Plant Science*, **165**(4), 803-810.

47   Aijun, M., Quigg, A., Schwehr, K., Xu, C. and Santschi, P. (2007). Engineered silver nanoparticles in coastal marine environments: Bioavailability and toxic effects to phytoplankton *Thalassiosira weissflogii*. Abstract 1.10. *Nanoparticles and nanomaterials, 2nd International Meeting*, September 2007, Natural History Museum, London.

48   Handy, R.D., von der Kammer, F., Lead, J.R., Hassellov, M., Owen, R. and Crane, M. (2008). The ecotoxicology and chemistry of manufactured nanoparticles. *Ecotoxicology*, **17**, 287-314.

49   Oberdörster, E., Zhu, S., Blickley, T., McClellan-Green, P. and Haasch, M. (2006). Ecotoxicology of carbon based engineered nanoparticles: Effects of fullerenes (C_{60}) on aquatic organisms. *Carbon* **44**, 1112-1120.

50   Royal Commission on Environmental Pollution (RCEP) (2004). *A Limited Report on Biomass as a Renewable Energy Source*. TSO, London. Available at: www.rcep.org.uk

51   Boxall, A.B.A., Tiede, K. and Chaudhry, Q. (2007b). Engineered nanomaterials in soils and water: How do they behave and could they pose a risk to human health? *Nanomedicine*, **2**, 919-927.

52   Lyon, D.Y., Thill, A., Rose, J. and Alvarez, P.J.J. (2007). Ecotoxicological impacts of nanomaterials. In: *Environmental Nanotechnology: Applications and Implications of Nanomaterials*. Editors: M.R. Weisner and J.Y. Bottero. McGraw-Hill, NY.

53   Photograph reproduced by kind permission of Napier University, Edinburgh.

54   Stone, V., Fernandes, T.F., Ford, A.T. and Christofi, N. (2006). Suggested strategies for the testing of nanoparticles. *Mater. Res. Soc. Symp. Proc.*, **895**, 0895-G04-03-S04-03.1-03.6.

55   Templeton, R.C., Ferguson, P.L., Washburn, K.M., Scrivens, W.A. and Chandler, G.T. (2006). Life-cycle effects of single-walled carbon nanotubes (SWCNTs) on an estuarine meiobenthic copepod. *Environ. Sci. Technol.*, **40**, 7387-7393.

56   Roberts, A.P., Mount, A.S., Seda, B., Souther, J., Qiao, R., Lin, S., Ke, P.C., Rao, A.M. and Klaine, S.J. (2007). *In vivo* biomodification of lipid-coated carbon nanotubes by *Daphnia magna*. *Environ. Sci. Technol.*, **41**, 3025-3029.

57   *Ibid.*

58   Oberdörster (2004).

59   Cheng, J., Flauaut, E. and Cyheng, S.H. (2007). Effect of carbon nanotubes on developing zebrafish (*Danio rerio*). *Environ. Toxicol. Chem.*, **26**, 708-716.

86

60   Kashiwada, S. (2006). Distribution of nanoparticles in the see-through medaka (*Oryzias latipes*). *Environ. Health Persp.*, **114**, 1697-1702.

61   Dubertret, B., Skourides, P., Norris, D.J., Noireaux, V., Brivanlou, A.H. and Libchaber, A. (2002). *In vivo* imaging of quantum dots encapsulated in phospholipid micelles. *Science*, **298**(5599), 1759-1762.

62   Zhang *et al.* (2007).

63   Lead *et al.* (2008).

64   Henry *et al.* (2007).

65   Moore, M.N. (2006). Do nanoparticles present ecotoxicological risks for the health of the aquatic environment? *Environ. International*, **32**, 967-976.

66   Colvin, V.L. (2003). The potential environmental impact of engineered nanomaterials. *Nat. Biotechnol.*, **21**, 1166-1170.

67   Oberdörster *et al.* (2006).

68   Holbrook *et al.* (2008).

69   Owen, R. and Depledge, M.H. (2005). Nanotechnology and the environment: Risks and rewards. *Marine Pollution Bulletin*, **50**(6), 609-612.

70   Personal communication with Dr Sam Luoma, Natural History Museum, London, June 2008.

71   Personal communication with Mark Chappell, US Army Corps of Engineers, and Chris Metcalfe, Trent University, Peterborough, Canada, April 2008.

72   Zhang, T., Stilwell, J.L., Gerion, D., Ding, L., Elboudwarej, O., Cooke, P.A., Gray, J.W., Alivisatos, A.P. and Chen, F.F. (2006). Cellular effect of high doses of silica coated quantum dot profiled with high throughput gene expression analysis and high content cellomics measurements. *Nano. Lett.*, **6**, 800-808.

73   Personal communication with Mark Chappell, US Army Corps of Engineers, and Chris Metcalfe, Trent University, Peterborough, Canada, April 2008.

74   RCEP (2003). *Twenty-fourth Report: Chemicals in Products: Safeguarding the Environment and Human Health*. TSO, London. Available at: www.rcep.org.uk

75   Helland *et al.* (2007).

76   Boxall *et al.* (2007a).

77   Baumann, B., Rolf, O., Jakob, F., Goebel, S., Sterner, T., Eulert, J. and Rader, C.P. (2006). Synergistic effects of mixed TiAlV and polyethylene wear particles on TNFalpha response in THP-1 macrophages. *Biomed. Tech. (Berl.)*, **51**(5-6), 360-366.

78   Guntur, V.P. and Dhand, R. (2007). Inhaled insulin: Extending the horizons of inhalation therapy. *Respir. Care*, **52**(7), 911-922.

79   Geiser, M., Casaulta, M., Kupferschmid, B., Schulz, H., Semmler-Behnke, M. and Kreyling, W. (2008). The role of macrophages in the clearance of inhaled ultrafine titanium dioxide particles. *Am. J. Respir. Cell. Mol. Biol.*, **38**(3), 371-376.

80   Baumann *et al.* (2006).

81   Oberdörster, G., Sharp, Z., Atudorei, V., Elder, A., Gelein, R., Kreyling, W. and Cox, C. (2004). Translocation of inhaled ultrafine particles to the brain. *Inhal. Toxicol.*, **16**(6-7), 437-445.

82   Donaldson, K. and Tran, C.L. (2002). Inflammation caused by particles and fibers. *Inhal. Toxicol.*, **14**(1), 5-27.

83   Bunn, H.J., Dinsdale, D., Smith, T. and Grigg, J. (2001). Ultrafine particles in alveolar macrophages from normal children. *Thorax*, **56**(12), 932-934.

84   Guntur and Dhand (2007).

85   Kreyling, W.G., Semmler, M., Erbe, F., Mayer, P., Takenaka, S., Schulz, H., Oberdörster, G. and Ziesenis, A. (2002). Translocation of ultrafine insoluble iridium particles from lung epithelium to extrapulmonary organs is size dependent but very low. *J. Toxicol. Environ. Health A*, **65**(20), 1513-1530.

86 Yacobi, N.R., Phuleria, H.C., Demaio, L., Liang, C.H., Peng, C.A., Sioutas, C., Borok, Z., Kim, K.J. and Crandall, E.D. (2007). Nanoparticle effects on rat alveolar epithelial cell monolayer barrier properties. *Toxicol. In Vitro,* **21**(8), 1373-1381.

87 Nemmar, A., Hoet, P.H., Vanquickenborne, B., Dinsdale, D., Thomeer, M., Hoylaerts, M.F., Vanbilloen, H., Mortelmans, L. and Nemery, B. (2002). Passage of inhaled particles into the blood circulation in humans. *Circulation,* **105**(4), 411-414;

Mills, N.L., Amin, N., Robinson, S.D., Anand, A., Davies, J., Patel, D., de la Fuente, J.M., Cassee, F.R., Boon, N.A., MacNee, W., Millar, A.M., Donaldson, K. and Newby, D.E. (2006). Do inhaled carbon nanoparticles translocate directly into the circulation in humans? *Am. J. Respir. Crit. Care Med.,* **173**(4), 426-431;

Wiebert, P., Sanchez-Crespo, A., Falk, R., Philipson, K., Lundin, A., Larsson, S., Möller, W., Kreyling, W.G. and Svartengren, M. (2006). No significant translocation of inhaled 35-nm carbon particles to the circulation in humans. *Inhal. Toxicol.,* **18**(10), 741-747.

88 Frampton, M.W., Utell, M.J., Zareba, W., Oberdörster, G., Cox, C., Huang, L.S., Morrow, P.E., Lee, F.E., Chalupa, D., Frasier, L.M., Speers, D.M. and Stewart, J. (2004). Effects of exposure to ultrafine carbon particles in healthy subjects and subjects with asthma. *Res. Rep. Health Eff. Inst.,* **December**(126), 1-47.

89 Elder, A., Gelein, R., Silva, V., Feikert, T., Opanashuk, L., Carter, J., Potter, R., Maynard, A., Ito, Y., Finkelstein, J. and Oberdörster, G. (2006). Translocation of inhaled ultrafine manganese oxide particles to the central nervous system. *Environ. Health Perspect.,* **114**(8), 1172-1178. Erratum in: *Environ. Health Perspect.,* **114**(8), 1178.

90 Yoo, J.W., Kim, Y.S., Lee, S.H., Lee, M.K., Roh, H.J., Jhun, B.H., Lee, C.H. and Kim, D.D. (2003). Serially passaged human nasal epithelial cell monolayer for *in vitro* drug transport studies. *Pharm. Res.,* **20**(10), 1690-1696.

91 Hillyer, J.F. and Albrecht, R.M. (2001). Gastrointestinal persorption and tissue distribution of differently sized colloidal gold nanoparticles. *J. Pharm. Sci.,* **90**(12), 1927-1936.

92 Nohynek, G.J., Lademann, J., Ribaud, C. and Roberts, M.S. (2007). Grey goo on the skin? Nanotechnology, cosmetic and sunscreen safety. *Crit. Rev. Toxicol.,* **37**(3), 251-277.

93 Tinkle, S.S., Antonini, J.M., Rich, B.A., Roberts, J.R., Salmen, R., DePree, K. and Adkins, E.J. (2003). Skin as a route of exposure and sensitization in chronic beryllium disease. *Environ. Health Perspect.,* **111**(9), 1202-1208.

94 Ryman-Rasmussen, J.P., Riviere, J.E. and Monteiro-Riviere, N.A. (2006). Penetration of intact skin by quantum dots with diverse physicochemical properties. *Toxicol. Sci.,* **91**(1), 159-165;

Ryman-Rasmussen, J.P., Riviere, J.E. and Monteiro-Riviere, N.A. (2007). Surface coatings determine cytotoxicity and irritation potential of quantum dot nanoparticles in epidermal keratinocytes. *J. Invest. Dermatol.,* **127**(1), 143-153.

95 McGrath, J.A. and Uitto, J. (2008). The filaggrin story: Novel insights into skin-barrier function and disease. *Trends Mol. Med.,* **14**(1), 20-27.

96 Holgate, S.T. (2007). Epithelium dysfunction in asthma. *J. Allergy Clin. Immunol.,* **120**(6), 1233-1244.

97 Maynard, A.D. and Kuempel, E.D. (2005). Airborne nanostructured particles and occupational health. *J. Nanoparticle Res.,* **7**, 587-614.

98 Oberdörster, G., Ferin, J. and Lehnert, B.E. (1994). Correlation between particle size, *in vivo* particle persistence, and lung injury. *Environ. Health Perspect.,* **102**(Suppl. 5), 173-179.

99 Kreyling, W.G., Semmler-Behnke, M. and Möller, W. (2006). Ultrafine particle-lung interactions: Does size matter? *J. Aerosol Med.,* **19**(1), 74-83.

100 Kaewamatawong, T., Shimada, A., Okajima, M., Inoue, H., Morita, T., Inoue, K. and Takano, H. (2006). Acute and subacute pulmonary toxicity of low dose of ultrafine colloidal silica particles in mice after intratracheal instillation. *Toxicol. Pathol.,* **34**(7), 958-965.

101 Lam, C.W., James, J.T., McCluskey, R., Arepalli, S. and Hunter, R.L. (2006). A review of carbon nanotube toxicity and assessment of potential occupational and environmental health risks. *Crit. Rev. Toxicol.,* **36**(3), 189-217.

102 Shvedova, A.A., Kisin, E.R., Mercer, R., Murray, A.R., Johnson, V.J., Potapovich, A.I., Tyurina, Y.Y., Gorelik, O., Arepalli, S., Schwegler-Berry, D., Hubbs, A.F., Antonini, J., Evans, D.E., Ku, B.K., Ramsey, D., Maynard, A., Kagan, V.E., Castranova, V. and Baron, P. (2005). Unusual inflammatory and fibrogenic pulmonary responses to single-walled carbon nanotubes in mice. *Am. J. Physiol. Lung Cell. Mol. Physiol.*, **289**, L698-L708.

103 Rödelsperger, K. (2004). Extrapolation of the carcinogenic potency of fibers from rats to humans. *Inhal. Toxicol.*, **16**(11-12), 801-807.

104 Poland, C.A., Duffin, R., Kinloch, I., Maynard, A., Wallace, W.A.H., Seaton, A., Stone, V., Brown, S., MacNee, W. and Donaldson, K. (2008). Carbon nanotubes introduced into the abdominal cavity of mice show asbestos-like pathogenicity in a pilot study. *Nature Nanotechnology*, **3**, 423-428.

105 *Ibid.*

106 Park, S., Lee, Y.K., Jung, M., Kim, K.H., Chung, N., Ahn, E.K., Lim, Y. and Lee, K.H. (2007). Cellular toxicity of various inhalable metal nanoparticles on human alveolar epithelial cells. *Inhal. Toxicol.*, **19**(Suppl. 1), 59-65.

107 Behnajady, M.A., Modirshahla, N., Shokri, M., Elham, H. and Zeininezhad, A. (2008). The effect of particle size and crystal structure of titanium dioxide nanoparticles on the photocatalytic properties. *J. Environ. Sci. Health A, Tox. Hazard. Subst. Environ. Eng.*, **43**(5), 460-746.

108 Warheit, D.B., Webb, T.R., Reed, K.L., Frerichs, S. and Sayes, C.M. (2007a). Pulmonary toxicity study in rats with three forms of ultrafine-TiO$_2$ particles: Differential responses related to surface properties. *Toxicology*, **230**(1), 90-104.

109 Monteiller, C., Tran, L., MacNee, W., Faux, S., Jones, A., Miller, B. and Donaldson, K. (2007). The pro-inflammatory effects of low-toxicity low-solubility particles, nanoparticles and fine particles, on epithelial cells *in vitro*: The role of surface area. *Occup. Environ. Med.*, **64**(9), 609-615.

110 Rothen-Rutishauser, B., Mühlfeld, C., Blank, F., Musso, C. and Gehr, P. (2007). Translocation of particles and inflammatory responses after exposure to fine particles and nanoparticles in an epithelial airway model. *Part. Fibre Toxicol.*, **4**, 9.

111 Limbach, L.K., Wick, P., Manser, P., Grass, R.N., Bruinink, A. and Stark, W.J. (2007). Exposure of engineered nanoparticles to human lung epithelial cells: Influence of chemical composition and catalytic activity on oxidative stress. *Environ. Sci. Technol.*, **41**(11), 4158-4163.

112 Kagan, V.E., Tyurina, Y.Y., Tyurin, V.A., Konduru, N.V., Potapovich, A.I., Osipov, A.N., Kisin, E.R., Schwegler-Berry, D., Mercer, R., Castranova, V. and Shvedova, A.A. (2006). Direct and indirect effects of single walled carbon nanotubes on RAW 264.7 macrophages: Role of iron. *Toxicol. Lett.*, **165**(1), 88-100.

113 Borm, P.J., Kelly, F., Künzli, N., Schins, R.P. and Donaldson, K. (2007). Oxidant generation by particulate matter: From biologically effective dose to a promising, novel metric. *Occup. Environ. Med.*, **64**(2), 73-74.

114 Oyelere, A.K., Chen, P.C., Huang, X., El-Sayed, I.H. and El-Sayed, M.A. (2007). Peptide-conjugated gold nanorods for nuclear targeting. *Bioconjug. Chem.*, **18**(5), 1490-1497.

115 Mroz, R.M., Schins, R.P., Li, H., Jimenez, L.A., Drost, E.M., Holownia, A., MacNee, W. and Donaldson, K. (2008). Nanoparticle-driven DNA damage mimics irradiation-related carcinogenesis pathways. *Eur. Respir. J.*, **31**, 241-251.

116 Twickler, M., Dallinga-Thie, G. and Cramer, M.J. (2006). Trojan horse hypothesis: Inhaled airborne particles, lipid bullets, and atherogenesis. *J. Am. Med. Assoc.*, **295**(20), 2354.

117 Sayes, C.M., Reed, K.L. and Warheit, D.B. (2007). Assessing toxicity of fine and nanoparticles: Comparing *in vitro* measurements to *in vivo* pulmonary toxicity profiles. *Toxicol. Sci.*, **97**(1), 163-180.

118 Shvedova, A.A., Fabisiak, J.P., Kisin, E.R., Murray, A., Roberts, J., Antonini, J., Kommineini, C., Reynolds, J., Barchwosky, A., Castranova, V. and Kagan, V.E. (2008). Combined exposure to carbon nanotubes and bacteria enhances pulmonary inflammation and infectivity. *Am. J. Respir. Cell. Mol. Biol.*, **38**, 579-590.

119 Fedulov, A.V., Leme, A., Yang, Z., Dahl, M., Lim, R., Mariani, T.J. and Kobzik, L. (2008). Pulmonary exposure to particles during pregnancy causes increased neonatal asthma susceptibility. *Am. J. Respir. Cell. Mol. Biol.*, **38**(1), 57-67.

120 Niwa, Y., Hiura, Y., Murayama, T., Yokode, M. and Iwai, N. (2007). Nano-sized carbon black exposure exacerbates atherosclerosis in LDL-receptor knockout mice. *Circ. J.*, **71**(7), 1157-1161.

121 Yatera, K., Hsieh, J., Hogg, J.C., Tranfield, E., Suzuki, H., Shih, C.H., Behzad, A.R., Vincent, R. and van Eeden, S.F. (2008). Particulate matter air pollution exposure promotes the recruitment of monocytes into atherosclerotic plaques. *Am. J. Physiol. Heart Circ. Physiol.*, **294**(2), H944-H953.

122 Evidence from the European Chemical Industry Council (CEFIC), July 2007; evidence from the Nanotechnology Industries Association (NIA), July 2007.

123 Evidence from Dr Andrew Maynard, Woodrow Wilson Center, July 2007; personal communication with Professor Vicki Colvin, Rice University, January 2008.

124 RCEP (2003).

125 RS/RAE (2004).

126 Personal communication with Professor Jim Bridges, University of Surrey and Chair of SCENIHR, June 2008.

127 Lead *et al.* (2008).

128 *Ibid.*

129 Personal communication with Professor Vicki Colvin, Rice University, January 2008.

130 *Research projects on the safety of nanomaterials: Reviewing the knowledge gaps.* European Commission Workshop, 17-18 April 2008, Brussels.

131 Warheit, D.B., Borm, P.J., Hennes, C. and Lademann, J. (2007b). Testing strategies to establish the safety of nanomaterials: Conclusions of an ECETOC workshop. *Inhal. Toxicol.*, **19**(8), 631-643.

132 Presentation by J. Steevens, US Army Corps of Engineers, at: *Risk, Uncertainty and Decision Analysis for Nanomaterials: Environmental Risks and Benefits and Emerging Consumer Products.* NATO-sponsored conference, 27-30 April 2008, Faro, Portugal.

133 Lead *et al.* (2008).

134 Stolpe, B., Hassellov, M., Andersson, K. and Turner, D.R. (2005). High resolution ICPMS as an on-line detector for flow field-flow fractionation; multi-element determination of colloidal size distributions in a natural water sample. *Analytica Chimica Acta*, **535**, 109-121.

135 Yohannes, G., Wiedmer, S.K., Hiidenhovi, J., Hietman, A. and Hyotylainen, T. (2007). Comprehensive two-dimensional field-flow fractionation-liquid chromatography in the analysis of large molecules. *Analytical Chemistry*, **70**, 3091-3098.

136 Presentation by J. Steevens, US Army Corps of Engineers, at: *Risk, Uncertainty and Decision Analysis for Nanomaterials: Environmental Risks and Benefits and Emerging Consumer Products.* NATO-sponsored conference, 27-30 April 2008, Faro, Portugal.

137 Hearing before the Committee on Science, House of Representatives. One Hundred Ninth Congress – Second Session (2006). *Research on Environmental and Safety Impacts of Nanotechnology: What are the Federal Agencies Doing?* Serial Number 109-63. Available at: www.house.gov/science

138 Evidence received from the Department for Innovation, Universities and Skills (DIUS), March 2008. Available at: www.rcep.org.uk

Chapter 4

1 Renn, O. and Roco, M. (2006). *White Paper on Nanotechnology Risk Governance.* International Risk Governance Council.

2 Royal Commission on Environmental Pollution (RCEP) (1976). *Sixth Report: Nuclear Power and the Environment.* HMSO, London.

3 NanoBio-RAISE Consortium (2008). *Ethical and Societal Issues in Nanobiotechnology.* 6th Framework Programme Science and Society Co-ordination Action. Paper in preparation.

4 European Environment Agency (EEA) (2001). *Late Lessons from Early Warnings: The precautionary principle 1896-2000.* Environmental Issue Report No. 22. EEA, Copenhagen.

5   Poland, C.A., Duffin, R., Kinloch, I., Maynard, A., Wallace, W.A.H., Seaton, A., Stone, V., Brown, S., MacNee, W. and Donaldson, K. (2008). Carbon nanotubes introduced into the abdominal cavity of mice show asbestos-like pathogenicity in a pilot study. *Nature Nanotechnology*, **3**, 423-428.

6   Personal communication with Professor Pedro Alvarez, Rice University, January 2008.

7   Personal communication with Dr. Jamie Lead, University of Birmingham, June 2008.

8   Barber, B. (1983). *The Logic and Limits of Trust*. Rutgers University Press, New Brunswick, NJ.

9   Personal communication with Ministry of Economy, Trade and Industry (METI), Japan, November 2007.

10   Personal communication with US Food and Drug Administration (FDA), January 2008, and US Environmental Protection Agency (EPA), January 2008.

11   Brickman, R., Jasanoff, S. and Ilgen, T. (1983). *Controlling Chemicals: The Politics of Regulation in Britain and the United States*. Cornell University Press, Ithaca, NY;
     Jasanoff, S. (2005). *Designs on Nature: Science and Democracy in Europe and the United States*. Princeton University Press, Princeton, NJ.

12   Funtowicz, S. and Ravetz, J. (1985). Three kinds of risk assessment: A methodological analysis. In: *Risk Analysis in the Private Sector*. Editors: C. Whipple and V. Covello. Plenum Press, New York.

13   Regulation (EC) No. 1907/2006 of the European Parliament and of the Council of 18 December 2006 concerning the Registration, Evaluation, Authorisation and Restriction of Chemicals (REACH), establishing a European Chemicals Agency, amending Directive 1999/45/EC and repealing Council Regulation (EEC) No. 793/93 and Commission Regulation (EC) No. 1488/94 as well as Council Directive 76/769/93 and Commission Directives 91/155/EEC, 93/67/EEC, 93/105/EC and 2000/21/EC. *Official Journal of the European Union*, **L396**, 30 December 2006.

14   Directive 2002/96/EC of the European Parliament and of the Council of 27 January 2003 on Waste Electrical and Electronic Equipment (WEEE). *Official Journal of the European Union*, **L37**, 13 February 2003.

15   Applegate, J.S. (2008). *Synthesising TSCA and Reach: Practical Principles for Chemical Regulation Reform*. ExpressO. Available at: http://works.bepress.com/john_applegate/1. Accessed 22 July 2008;
     Molyneux, C.G. (2008). *Chemicals. The Yearbook of European Environmental Law. Volume 8*. Editors: T. Etty and H. Somsen. ISBN: 978019954261. In press.

16   Molyneux (2008).

17   *The ENDS Report*, **Issue 401**, June 2008.

18   Chaudhry, Q., Blackburn, J., Floyd, P., George, C., Nwaogu, T., Boxall, A. and Aitken, R. (2006). *A scoping study to identify gaps in environmental regulation for the products and applications of nanotechnologies. Final report*. Report by the Central Science Laboratory for the Department for Environment, Food and Rural Affairs (Defra). Available at: http://www.defra.gov.uk/science/Project_Data/DocumentLibrary/CB01075/CB01075_3373_FRP.doc. Accessed 29 July 2008.

19   Frater, L., Stokes, E., Lee, R. and Oriola, T. (2006). *An overview of the framework of current regulation affecting the development and marketing of nanomaterials*. A report for the Department for Trade and Industry (DTI). December 2006.

20   *Ibid.*

21   European Commission Health and Consumer Protection Directorate General (DG SANCO) (2004). *Nanotechnologies: A preliminary risk analysis on the basis of a workshop organised in Brussels on 1-2 March 2004*. Available at: http://ec.europa.eu/health/ph_risk/documents/ev_20040301_en.pdf. Accessed 28 July 2008.

22   UK Government (2008). *Statement about Nanotechnologies*. Available at: http://www.dius.gov.uk/policy/documents/statement-nanotechnologies.pdf. Accessed 23 July 2008.

23 Chaudhry *et al.* (2006);

Food Standards Agency (FSA) (2006). *A review of potential implications of nanotechnologies for regulations and risk assessment in relation to food.* FSA, London;

Frater *et al.* (2006);

Health and Safety Executive (HSE) (2006). *Review of the Adequacy of Current Regulatory Regimes to Secure Effective Regulation of Nanoparticles Created by Nanotechnology. The Regulations covered by HSE.* Available at: http://www.hse.gov.uk/horizons/nanotech/regulatoryreview.pdf. Accessed 23 July 2008.

24 Personal communication with European Commission officials, February 2008, and Defra, March 2008.

25 European Commission Scientific Committee on Emerging and Newly Identified Health Risks (SCENIHR) (2008). *Request for Scientific Opinion on Risk Assessment of Products of Nanotechnologies.* Available at: http://ec.europa.cu/health/ph_risk/committees/04_scenihr/docs/scenihr_q_015.pdf. Accessed 23 July 2008.

26 *The ENDS Report,* **Issue 401**, June 2008.

27 European Commission (2007a). *Questions and Answers on REACH.* Updated version from February 2007. Available at: http://ec.europa.eu/environment/chemicals/reach/pdf/qa_july07.pdf. Accessed 23 July 2008.

28 Personal communication with Professor Vicki Colvin, Rice University, January 2008, and Dr Andrew Maynard, Woodrow Wilson Center, January 2008.

29 Personal communication with US EPA, January 2008.

30 Presentation to RCEP by Professor Robert Maynard, Health Protection Agency, March 2007.

31 Arnall, H.W. (2003). *Future Technologies, Today's Choices. Nanotechnology, Artificial Intelligence and Robotics. A technical, political and institutional map of emerging technologies.* Greenpeace Environmental Trust UK.

32 Royal Society and Royal Academy of Engineering (RS/RAE) (2004). *Nanoscience and Nanotechnologies: Opportunities and uncertainties.* Royal Society Policy Document 19/04.

33 Personal communication with US FDA, January 2008.

34 European Commission (2008). *Code of Conduct for Responsible Nanosciences and Nanotechnologies Research.* Available at: http://ec.europa.eu/nanotechnology/index_en.html. Accessed 23 July 2008.

35 Royal Society, Insight Investment, Nanotechnology Industries Association (2007). *Code of Conduct for Responsible Nanotechnology.* Available at: http://www.responsiblenanocode.org/. Accessed 23 July 2008.

36 Oral evidence from Defra, 2008.

37 *The Hazardous Waste (England and Wales) Regulations 2005.* Statutory Instrument 2005 No. 894. TSO, London.

38 Oral evidence from Defra, 2008.

39 Defra (2008). *Voluntary Reporting Scheme for Engineered Nanoscale Materials.* Available at: http://defraweb/environment/nanotech/policy/index.htm. Accessed 23 July 2008.

40 Oral evidence from Defra, 2008.

41 Cambridge Water Company vs. Eastern Counties Leather plc (1994). 2 AC 264.

42 RCEP (2003). *Twenty-fourth Report: Chemicals in Products: Safeguarding the Environment and Human Health.* TSO, London. Available at: http://www.rcep.org.uk;

RCEP (2005). *Crop Spraying and the Health of Residents and Bystanders.* TSO, London.

43 Personal communication with the American Chemistry Council, January 2008.

44 Wilsdon, J. and Willis, R. (2004). *See-through science: Why public engagement needs to move upstream.* Demos, London;

Macnaghten, P., Kearnes, M.B. and Wynne, B. (2005). Nanotechology, governance and public deliberation: What role for the social sciences? *Science Communication,* **27**, 258-291;

Stirling, A. (2005). Opening up or closing down? Analysis, participation and power in the social appraisal of technology. In: *Science, Citizenship and Globalisation.* Editors: M. Leach, I. Scoones and B. Wynne. Zed, London.

45 European Commission (2007b). *Taking European Knowledge Society Seriously.* Report of the Expert Group on Science and Governance to the Science, Economy and Society Directorate, Directorate General for Research, European Commission. Chairman: Brian Wynne. EUR 22700. Brussels.

46 *Ibid.*

47   Gaskell, G., Allum, N.C. and Stares, S. (2003). *Europeans and Biotechnology in 2002*. Eurobarometer 58.0. Methodology Institute, London School of Economics;

BMRB Social Research (2004). *Nanotechnology: Views of the general public*. Quantitative and qualitative research carried out as part of the Nanotechnology Study. Prepared for the RS/RAE Nanotechnology Working Group. London;

Cobb, M.D. and Macoubrie, J. (2004). Public perceptions about nanotechnology: Risks, benefits, trust. *Journal of Nanoparticle Research*, **6**(4), 395-405;

Macoubrie, J. (2005). *Informed Public Perceptions of Nanotechnology and Trust in Government. Project on Emerging Nanotechnologies*. Woodrow Wilson Center, Washington DC.

48   Doubleday, R. (2007). Risk, public engagement and reflexivity: Alternative framings of the public dimensions of nanotechnology. *Health, Risk & Society* **9**(2), 1-17;

presentation by J. Steevens, US Army Corps of Engineers, at: *Risk, Uncertainty and Decision Analysis for Nanomaterials: Environmental Risks and Benefits and Emerging Consumer Products*. NATO-sponsored Conference, 27-30 April 2008, Faro, Portugal.

49   RS/RAE (2004).

50   Meridian Institute (2003). *Global Dialogue on Nanotechnology and the Poor: Opportunities and Risks*. Available at: http://www.meridian-nano.org. Accessed 28 July 2008;

Nanojury (2005). Available at: http://www.nanojury.org.uk/index.html. Accessed 28 July 2008;

Gavelin, K. and Wilson, R. (2006). *Democratic Technologies?* The final report of the Nanotechnology Engagement Group. Involve, London;

Smallman, M. and Nieman, A. (2006). *Small Talk: Discussing Nanotechnologies*. Think-Lab Ltd, London. Available at: http://www.smalltalk.org.uk/. Accessed 28 July 2008;

Stilgoe, J. (2007). *Nanodialogues: Experiments in public engagement with science*. Demos, London. Available at: http://www.demos.co.uk/projects/thenanodialogues/overview. Accessed 28 July 2008;

Wuppertal Institute for Climate, Environment and Energy, EMPA, Forum for the Future and Triple Innova (2007). *The Nanologue Project: Europe-wide dialogue on social, ethical and legal impacts of nanotechnology*. European Union 6th Framework Programme. Available at: http://www.nanologue.net/index.php?seite=2. Accessed 28 July 2008.

51   Doubleday (2007).

52   Kearnes, M., Macnaghten, P. and Wilsdon, J. (2006). *Governing at the Nanoscale: People, Policies and Emerging Technologies*. Demos, London.

53   Personal communication with the Ministry of Education, Culture, Sports, Science and Technology (MEXT), Japan, November 2007.

54   HM Government (2005). *The Government's Outline Programme for Public Engagement on Nanotechnologies*. DTI, London.

55   Gavelin and Wilson (2006).

56   Engineering and Physical Sciences Research Council (EPSRC) (2008). *Nanoscience through Engineering to Application. Third Grand Challenge – Consultation*. Available at: http://www.epsrc.ac.uk/ResearchFunding/Programmes/Nano/RC/GrandChallengesNanotechnology.htm. Accessed 23 July 2008.

57   Rodemeyer, M., Sarewitz, D. and Wilsdon, J. (2005). *The Future of Technology Assessment*. Woodrow Wilson Center, Washington DC. Available at: http://www.wilsoncenter.org/news/docs/techassessment.pdf. Accessed 28 July 2008.

58   Rip, A., Misa, T.J. and Schot, J. (Eds.) (1995). *Managing Technology in Society: The Approach of Constructive Technology Assessment*. Pinter, London.

59   Personal communication with Professor Phil Macnaghten, University of Durham, June 2008.

60   *Ibid*.

93

61   International Risk Governance Council (IRGC) (2006). *Risk Governance: Towards an Integrative Approach.* IRGC, Geneva;

Pidgeon, N. and Rogers-Hayden, T. (2007). Opening up nanotechnology dialogue with the publics: Risk communication or upstream engagement. *Health, Risk & Society,* **9**(2), 191-210.

62   Wilsdon, J., Wynne, B. and Stilgoe, J. (2005). *The Public Value of Science: Or How to Ensure that Science Really Matters.* Demos, London.

63   Wilsdon and Willis (2004);

Stirling (2005).

64   Jasanoff (2005).

Appendix A

ANNOUNCEMENT OF THE STUDY AND INVITATION TO SUBMIT EVIDENCE

A1 ANNOUNCEMENT OF THE STUDY

The Royal Commission study on novel materials in the environment was announced in a news release on 3 April 2006 in the following terms. Over 100 organisations were invited to respond to issues described below and around 20 responses were received.

ROYAL COMMISSION STUDY ON NOVEL MATERIALS IN THE ENVIRONMENT

The next study by the Royal Commission on Environmental Pollution will be on the environmental effects of novel materials and applications.

Novel materials, along with new forms and applications of existing chemicals, are continually being developed to help make technological advances and improve performance, driven by the needs of industry and the demands of society. Although much attention has been paid to the effects of the environment on novel materials and their behaviour in different situations, there is relatively little work on the environmental effects of novel materials. However, governments are starting to fund research programmes and develop policy in this area and our study will contribute to that process.

The Commission will look at all aspects of the environmental effects of novel materials and applications, including both benefits and potentially harmful impacts.

The environmental effects of novel materials and applications was selected as a topic for study after consultation on the Commission's future work programme. The study will draw on approaches used in previous reports by the Commission, including the *13th report: The release of genetically engineered organisms to the environment*, and *24th report: Chemicals in products*.

The study is likely to cover a number of themes, including for example, the development process of novel materials, toxicity and ecotoxicity testing and data, positive and negative environmental impacts, and whether novel materials and applications are adequately regulated under existing regulations.

BACKGROUND TO THE STUDY

Novel materials, along with new forms and applications of existing chemicals are continually being developed to help make technological advances and improve performance, mainly in the fields of engineering and IT, but also in many other fields. An example of such a development is rhenium, which has previously been just a waste product from copper mining. It is now used in nickel alloys for jet engines, enabling them to fly at temperatures about fifty degrees centigrade higher than previously, so lowering fuel consumption.

Nanotechnology and nanoscience are also developing at a rapid pace. Current uses include sunscreens based around microfine particles, car bumpers made from nanocomposites[i] and coatings made from titanium dioxide nanoparticles to produce self-cleaning windows.

Lately, governments have started to look into this issue, developing policies and funding research. The majority of work carried out in this field has been on nanoscience and technology. The Royal Society in collaboration with the Royal Academy of Engineering published a policy document called *Nanoscience and Nanotechnologies: Opportunities and uncertainties* in July 2004. The report was wide ranging and included a section on the environmental effects of nanoscience and technology. The UK Government published its response to that report in February 2005, and agreed that further research on environmental effects would need to be carried out.

The Office of Science and Technology has set up the inter-governmental Nanotechnology Issues Dialogue Group (NIDG) which will co-ordinate Government activities in this field, and provide evidence to inform the Council for Science and Technology's 2- and 5-year reviews of Government's progress on this issue.

The Department for Environment, Food and Rural Affairs (Defra) is also looking at the environmental effects of these new technologies, using the Royal Society/Royal Academy of Engineering report as a basis. The Advisory Committee on Hazardous Chemicals has received a number of presentations on the subject. The European Commission also published a 4-year action plan on nanotechnology in June 2005.

The environmental impacts of other new materials, such as rare earth metals in electronic components, in use or in development appear to be less well studied.

The study will be addressing UK policies and programmes and will make recommendations to the UK Government, but the Commission will also look at work being carried out at the EU and global level.

BROAD TOPICS TO BE COVERED

Novel materials and applications cover a wide range of scientific, engineering and technological fields. There are a number of possible ways to subdivide this topic into categories for investigation and how to do this for the purposes of the Royal Commission's report will be one of the first issues that will have to be addressed. An example of this is demonstrated by the European Commission which has divided the field into four for the purposes of its research programme, including:

Crosscutting materials technologies: This involves developing novel materials with wide ranging application potential, and includes nanotechnology, surface engineering and materials processing technologies;

Advanced functional materials: This involves highly advanced materials with multi-sector use, including electronics, magnetic/optical materials, sensors and industrial systems and biomaterials;

---

i   Department for Environment, Food and Rural Affairs (Defra) Science Notes: *Nanoscience and the Environment.*

<u>Sustainable chemistry</u>: This covers the development of sustainable industrial chemistry with efficient use of resources and recycled materials, such as chemical engineering, advanced chemical reactions and chemistry for new materials; and

<u>Structural materials</u>: This covers all types of engineering.[ii]

As novel and advanced materials and applications are released into industrial processes and the marketplace, they will be affected by, and have effects on the environment. The expansion of work in this area and the raising of its profile have meant increased interest and awareness in the subject. It is the Royal Commission's intent to make a wide-ranging investigation, looking at different categories of novel materials and applications, including nanomaterials, positive and negative environmental impacts of novel materials, risk assessment and management, the regulatory framework and the identification of research gaps.

Broad topics that might be covered include:

- the development process of new materials;

- the life cycle analysis of these materials;

- toxicity and ecotoxicity issues;

- what the potential impacts on human health in terms of environmental exposure are;

- what the potential environmental impacts are, both positive and negative, along with possible ways of dealing with them;

- whether novel materials and applications are adequately regulated under existing environmental regulations; and

- waste issues: some products containing novel materials have a short lifespan and may not be recyclable.

The breadth of this study is potentially very wide, depending on the definition of novel materials used. Therefore, the Commission is not minded to investigate the use of GM technology, nor the human health aspects of pharmaceuticals or medical devices.

A2 INVITATION TO SUBMIT EVIDENCE

After considering the responses to the original announcement, the Royal Commission wrote to over 250 organisations and individuals in April 2007 for evidence on the following questions. Around 50 responses were received.

ROYAL COMMISSION STUDY ON NOVEL MATERIALS IN THE ENVIRONMENT – INVITATION TO SUBMIT EVIDENCE

The study is investigating the environmental effects of novel materials and novel applications of existing materials. It will provide an authoritative framework for thinking about and addressing the impacts of

ii   Taken from the website of the European Commission Research Directorate General: http://europa.eu.int/comm/research/growth/gcc/ga01.html#top

such materials. To help with this task, the Commission is keen to hear the views of organisations and individuals with an interest in novel materials and applications.

You will see that the scope of the study relates to novel materials and their applications. There is some discussion of this in relation to the first question posed. Our aim is to include not only nanotechnology, but also to cover other areas. We are aware that there are other significant areas of innovation and that many elements, which have largely been restricted to the academic laboratory, or older materials now finding new applications, are now entering products and these will ultimately lead to potential exposures of the wider environment and the general population. We therefore wish to consider these issues as well as those specific to nanotechnology in our work.

Issues on which the Commission would welcome evidence

The questions below are not intended to limit the Commission's study, but rather to highlight areas where Members believe they are most in need of input at this stage. You do not need to address all the issues listed. Indeed, you may wish to provide evidence on only a few. The list of questions occasionally provides, in italics, a commentary which reflects our initial thinking, designed primarily to illicit a response from consultees rather than to suggest that we have already closed down on our thinking. We would be pleased to know whether you agree with our initial thinking and if not, where you differ and why.

Study on the Environmental Effects of Novel Materials and Applications – Questions for Written Evidence Exercise

Novel materials, along with new forms and applications of existing chemicals are continually being developed to help make technological advances and improve performance. An example of such a development is rhenium, which has previously been just a waste product from copper mining. It is now used in nickel alloys for jet engines, enabling them to fly at temperatures significantly higher than previously, so lowering fuel consumption. Nanotechnology and nanoscience are also developing at a rapid pace.

Although there is a large body of work which looks at the effects of the environment on novel materials, there are very few studies on the environmental impacts of novel materials. The study could therefore be usefully broken down into three broad themes:

- Scene-setting: What are novel materials and what developments are likely over the next 5-10 years? Which ones should be investigated for the purposes of the study?

- Environmental and health impacts of novel materials.

- Governance and regulation issues.

Theme 1: Scene-setting: What are novel materials and what developments are likely over the next 5-10 years? Which ones should be investigated for the purposes of the study?

1   **What do you understand by the term novel material? How might novel materials best be classified? What novel materials should be included in the study?**

We have deliberately framed our inquiry to extend beyond nanotechnology per se. However, we do not intend to address all innovation and in particular, do not feel that it would be appropriate to cover the large number of organic molecules introduced each year, particularly those produced in small quantities in highly regulated and specialist sectors. This includes the pharmaceutical and biocide industries where the biological activity at least (though not the full spectrum of potential environmental effects) is characterised as part of the product development and approval process. In our view the study should be based on a working definition of novel materials which would encompass:

- *New uses for existing materials where the new usage may lead to substantially different exposures and hazards than current uses so that experience based on the current usage may not be a good indication of potential problems.*

- *New forms of existing materials: this is intended to include nanomaterials where significantly different functionalities are developed as a result of changing the scale and shape and arrangement of the particles at the nanolevel, e.g. the expression of significant chemical activity at the nanoscale of materials such as noble metals which may exhibit significant chemical reactivity or biocidal effects which are not manifested in the bulk form.*

- *Use of new materials such as metallic elements (rhodium, yttrium, etc.) and compounds derived from them. There is likely to be some interplay between all these categories, for example some nanotechnology products will include not only new forms of existing materials, but also other new substances, either as adjuncts, dopants or ligands and both new materials and nanoproducts may both be ultimately incorporated in new devices such as ICT equipment.*

2   At what point does a novel material cease to be novel?

What lies behind this question is to some extent an exploration of the degree to which unexpected environmental consequences may not emerge until some considerable time after the material has been in widespread use. The effect of refrigerant gases on the ozone layer is clearly one example but the developing evidence about the long-term effects of non-degradable plastics particularly in marine and aquatic environments may well be another. The issue also relates to questions of what mechanisms might be set in place to monitor environmental impacts to give warning (even if not always early warning) of potential difficulties as materials enter the environment and, even if they do not degrade, change in form through weathering and mechanical break-up into smaller particles over time.

3   What sort of materials and technologies are being developed – over the next 2, 5 and 10 years?

4   What are the drivers for the development of novel materials? What are the potential benefits of novel materials and the drivers for these?

What we have in mind here is essentially to try to tease out the underlying functionalities and products or improved performance which the use of new materials is seeking to deliver for society. Therefore it is part of the process of trying to identify the potential benefits which are discussed in more detail in Theme 2 below.

5   Can the development of novel materials have an impact on resource depletion?

6   Are issues of reuse and recycling considered when developing novel materials – e.g. could the phasing out of metals for composites make recycling difficult?

7   Are novel materials likely to alter the amount of waste generated and the ways in which it has to be handled?

Theme 2: Environmental and health impacts of novel materials

8   What are the most important impacts that novel materials could potentially have on the environment and human health? What are the main mechanisms and pathways for those impacts? How do we begin to conceptualise environmental impacts when we are in such unknown territory?

Embedded in this question are several issues. On the one hand there are fairly straightforward issues related to potential negative impacts through the biological effects of new materials on organisms (plants, animals and micro-organisms) in soil and water. Beyond this there is also the question of chemical and other interactions with parts of the environment such as the depletion of the ozone layer. There are also potential positive impacts where the use of the new material may allow the replacement of existing technologies which have significant negative impact on the environment. In addition, potential uses also exist for new materials in remediation and improvement of water and soil quality and improvements in the efficiency of processes such as energy generation and power transmission. New materials may also have indirect effects on the environment. For example, certain materials may be able to mobilise substances in soil in advantageous ways, but they could also lead to the mobilisation of hazardous material. These are intended only as examples and we would be very grateful for further thoughts on these issues.

9   Do novel materials have the potential to help 'solve' environmental problems, e.g. land contamination, energy generation? If so, how, and are there potential risks?

10   Do we have sufficient research and monitoring in terms of understanding toxicology and exposure in place in order to understand the effects of novel materials on the environment and human health?

11   Are current testing protocols 'fit for purpose' to test the potential environmental and health impacts of novel materials? If not, what needs to be developed or are there other strategies needed to address this issue?

12   Do we have adequate methodologies and instrumentation to detect and monitor engineered free nanoparticles in the environment?

13   Are the full life cycle impacts of novel materials being considered in terms of their potential effects on the environment and human health?

We are particularly concerned here about potential exposures through manufacture, use and disposal both in relation to the regulated official disposal routes (for instance for electronic products) and illegal or accidental losses, e.g. leakage from accidents, disposal direct to land/water. There are also issues about products which constantly abrade during use, creating dust or other mobile forms of release into the environment, even for products which are manufactured to be essentially fixed but may wear away in use.

14   How can you look at the effects of novel materials as a coherent whole, if they are even more difficult to categorise than nanomaterials?

15   Are there lessons to be learned from 'green chemistry' – and ways that manufacturing could be made more benign?

Theme 3: How to manage novel materials in society: Governance and regulation

16   **Is REACH the right framework for regulating novel materials and nanotechnologies?**

17   **Are the regulations which affect novel materials fit for purpose? Is existing legislation sufficient to deal with potential problems that could arise during the different stages of the novel material's life cycle, i.e. manufacture, use and disposal?**

18   **Is the UK, EU and global science and knowledge base sufficient to support current legislation frameworks and any future regulation? Where are the gaps and what are the research priorities?**

19   **Is the UK's and EU's research funding sufficient in this area? Is it being delivered in the right way?**

20   **Can novel materials and technologies be effectively governed and regulated if it is not possible to obtain exposure data before products containing novel materials are produced and made available to consumers?**

We have been made aware that even within carbon nanotubes there are potentially at least 10,000 possible formulations due to variations in substances added to the tubes and the actual physical size of the tubes themselves: all of these can affect functionality and potentially their environmental and biological behaviours. It is clearly not possible to apply conventional testing protocols because of the sheer numbers of formulations involved. If the industry is to develop, it is inevitable that there will be a degree of uncertainty. Although research can seek, and is seeking, to derive certain basic parameters to help identify and predict which materials may be problematic, some degree of uncertainty and ignorance is likely to remain. The social need is to develop regulatory mechanisms which reduce the risk of deleterious outcomes, while permitting the process of innovation to develop new materials for social benefit. The precautionary principle in its various formulations has been seen as one possible approach; would it be appropriate in this case or are there other approaches which would be preferable?

21   What is the role for engaging the range of different interests and perspectives, commercial, political, public and societal, on the development of novel materials in the context of global markets?

22   Are there general lessons to be learned from the development and use of other novel technologies, e.g. the development of genetically modified organisms?

23   How can an appropriate balance be achieved in the design of regulatory systems to effectively manage uncertainty?

24   What are the implications for liability when problems arise even if procedures are properly followed in good faith? Who should bear responsibility and what issues arise for insurance and redress?

25   How would you apply the precautionary principle to the management and regulation of novel materials?

26   In debate about new technologies, questions of need and control, as well as questions about consequences, have emerged as being important. To what extent should our study engage with questions about the need for novel and novel uses of materials; about who exercises control over such technologies; and about public trust in the institutions involved?

And finally:

27   Are there any other major questions or issues that the Commission should examine?

Appendix B

Conduct of the study

In order to carry out this study, the Royal Commission sought written and oral evidence, commissioned studies and advice on specific topics and made a number of visits.

Evidence

In parallel with the invitation to submit written evidence, which is reproduced in appendix A, the Secretariat wrote directly to a number of organisations and individuals.

The organisations and individuals listed below either submitted evidence or provided information on request for the purposes of the study, or otherwise gave assistance. In some cases, indicated by an asterisk, meetings were held with Commission Members or the Secretariat so that oral evidence could be given or particular issues discussed.

GOVERNMENT DEPARTMENTS

Department for Business, Enterprise and Regulatory Reform (BERR)*

Department for Environment, Food and Rural Affairs (Defra)*

Department of Health (DH)

Department for Innovation, Universities and Skills (DIUS)*

Ministry of Defence (MOD)

DEVOLVED ADMINISTRATIONS

Department of Environment Northern Ireland*

Scottish Executive

EUROPEAN AND INTERNATIONAL BODIES

Committee on Science and Technology, US House of Representatives*

European Centre for Ecotoxicology and Toxicology of Chemicals (ECETOC)

European Chemical Industry Council (CEFIC)

European Commission*

European Nanotechnology Trade Alliance (ENTA)*

Ministry of Economy, Trade and Industry, Japan (METI)*

Ministry of Education, Culture, Sports, Science and Technology, Japan (MEXT)*

Ministry of Health, Labour and Welfare, Japan (MHLW)*

National Institute for Materials Science, Japan (NIMS)*

National Institute for Occupational Safety and Health (NIOSH)

National Institute of Health Sciences, Japan (NIHS)*

National Science Foundation (NSF)*

Organisation for Economic Co-operation and Development (OECD)*

The National Academies

US Environmental Protection Agency*

US Food and Drug Administration*

Woodrow Wilson Center*

OTHER ORGANISATIONS

Advisory Committee on Hazardous Substances*

American Chemistry Council*

British Consulate-General, Boston

British Consulate-General, Houston

British Embassy, Tokyo

British Embassy, Washington DC

British Standards Institution

Centre for Business Relationships, Accountability, Sustainability and Society (BRASS), Cardiff University*

Chemical Industries Association

Cranfield University

Environment Agency*

Environmental Defense*

Forum for the Future

Green Chemistry Network

Greenpeace UK*

Health and Safety Executive

Hewlett Packard

Human Fertilisation and Embryology Authority (HFEA)*

Institute of Materials, Minerals and Mining

Marks and Spencer plc*

Nanotechnologies Stakeholders Forum*

Nanotechnology Industries Association (NIA)*

Nanotechnology Issues Dialogue Group*

Nokia*

Oxonica*

QinetiQ

Rolls Royce plc

Royal Society

Royal Society of Chemistry

Royal Society of Edinburgh*

Science and Innovation Network, Foreign and Commonwealth Office

Scottish Environment Protection Agency (SEPA)

Teijin*

Toshiba Corporation*

UK Environmental Law Association

UK Government Nanotechnologies Environmental Risk Assessment Task Force (Nanotechnology Research Co-ordination Group)

UK Research Councils

Which?

INDIVIDUALS

Dr Sirwan Arepalli*
Professor Pedro Alvarez*
Professor Paul Attfield*
Mr Mike Barry*
Professor Paul Bellamy
Professor Sir John Beringer*
Mr Cornelis Brekelmans*
Professor Jim Bridges
Dr Anthony Byrne*
Professor Roland Clift*
Professor Vicki Colvin*
Professor Peter Dobson*
Professor Ken Donaldson*
Dr Rob Doubleday
Professor Mauro Ferrari*
Professor Robert Flynn
Dr John Fortner*
Professor Derek Fray*
Dr Steffi Friedrichs
Dr Neil Glover*
Professor Peter Gregson*
Dr Richard Handy
Professor Ivan Holubek
Professor Vyvyan Howard
Professor Colin Humphreys
Professor Geoffrey Hunt
Professor Sheila Jasanoff*
Professor Richard Jones
Professor John Kilner*
Professor Neal Lane*
Dr Jamie Lead*
Dr Sam Luoma
Professor Philip Macnaghten
Dr Andrew Maynard*
Professor Robert Maynard*
Dr Celia Merzbacher*
Dr Julia Moore*
Professor Michael Moore
Professor Pierluigi Nicotera

Ms Ilga Nielsen
Professor Gunter Oberdörster*
Dr Richard Owen*
Dr Barry Park*
Dr Cathy Phillips*
Professor Nick Pidgeon
Professor Jane Plant*
Professor Jeremy Ramsden
Dr Jerome Ravetz
Dr Paul Reip*
Dr David Rejeski*
Professor Arie Rip*
Professor Robin Rogers
Professor Daniel Sarewitz*
Professor David Schiffrin
Professor Joanne Scott
Professor Anthony Seaton CBE*
Professor George Smith FRS*
Professor Mark Spearing
Mr Del Stark*
Professor Andrew Stirling*
Professor Vicki Stone*
Dr Clayton Teague*
Professor James M. Tour*
Dr Eva Valsami-Jones
Professor Lon Wilson*
Professor Anthony Walton
Professor Mark Welland*

COMMISSIONED STUDIES

A literature review was commissioned for the report during 2007 to provide information on the ecotoxicology of nanomaterials:

Literature Overview of the Ecotoxicology of Manufactured Nanoparticles. What evidence is there of unintended biological harm? Professor Tamara Galloway, School of Biosciences, University of Exeter.

In addition, more detailed reports were commissioned as follows:

Nanomaterial Innovation Systems: Their Structure, Dynamics and Regulation. Dr Adrian Smith, Dr Paul Nightingale, Mr Patrick van Zwanenberg, Dr Ismael Rafols and Molly Morgan, Science and Technology Policy Research Unit (SPRU), Freeman Centre, University of Sussex.

Regulation and the Chemical Industry. Mariana Doria, PhD Student, University of Trento.

VISITS

During the course of the study, Members of the Commission and its Secretariat made a series of visits. The Secretariat is indebted to the Science and Innovation Network teams at the British Embassies in Tokyo and Washington DC and the British Consulates-General in Houston and Boston for the assistance received in organising relevant itineraries.

May 2007, Belfast Members met the Vice-Chancellor of Queen's University Belfast, a researcher from the University of Ulster, and researchers involved with novel materials at Queen's University Belfast. Members also met with the Permanent Secretary of the Department of Environment Northern Ireland over dinner.

July 2007, Oxford Members visited the Oxford University Begbroke Science Park for a meeting with the Director, Professor Peter Dobson, and Dr Barry Park of Oxonica. Members were given presentations on nanotechnology research programmes at Oxford University and shown examples of products being developed which contain nanomaterials.

October 2007, Edinburgh Members visited the University of Edinburgh/Medical Research Council Centre for Inflammation Research based at the Queen's Medical Research Institute to meet Professor Ken Donaldson, Professor of Respiratory Toxicology, and Professor Vicki Stone from Napier University. The purpose of the visit was to gain a clearer insight into the issues of toxicology associated with nanoparticles.

November 2007, Derby Members visited Rolls Royce to attend a series of exhibitions and see their learning centre.

November 2007, Hampshire Members visited Qinetiq to attend a series of presentations and see their nanomaterials laboratory.

November 2007, Japan A delegation from the Royal Commission visited Tokyo to gather information in support of the novel materials study. The delegation met with representatives from the Ministry of Education, Culture, Sports, Science and Technology (MEXT), the Ministry of Economy, Trade and

Industry (METI), the National Institute for Materials Science (NIMS), the National Institute of Health Sciences (NIHS), the Ministry of Health, Labour and Welfare (MHLW), the Toshiba Corporation and Teijin.

January 2008, United States A delegation from the Royal Commission visited Washington, Houston and Boston to gather information in support of the novel materials study. The delegation met with representatives from the Office of Science and Technology, the National Nanotechnology Coordination Office, the Environmental Protection Agency, the American Chemistry Council, Environmental Defense, the National Academies, the Woodrow Wilson Center, the Food and Drug Administration, the National Science Foundation, the Committee on Science and Technology, and Rice University.

February 2008, Brussels A delegation from the Royal Commission met officials from DG Environment, DG Enterprise and Industry, DG Research and DG Health and Consumer Protection in the European Commission to discuss the EU chemical regulation, REACH, and their views about environment, health and governance issues relating to novel materials.

Appendix C

Seminar: Novel materials and applications: How do we manage the emergence of new technologies in democratic society?

On 11 January 2007 the Royal Commission hosted a seminar at the Institute of Materials, Minerals and Mining in London to gather views from interested parties relevant to deciding the scope of the novel materials study. The seminar involved around 40 participants and included speakers from a variety of backgrounds addressing such topics as: development of novel materials; health and environmental effects of novel materials; and governance and regulation of novel materials. The seminar had the following programme:

Welcome and Introduction by RCEP Chair: PROFESSOR SIR JOHN LAWTON

Key-note speech: PROFESSOR ANDREW STIRLING, Science and Technology Policy Research Unit (SPRU), University of Sussex

Session 1: Development of Novel Materials

Chair: PROFESSOR JANET SPRENT

PROFESSOR MARK WELLAND, University of Cambridge

DR PAUL REIP, QinetiQ Nanomaterials Ltd

Discussion

DR NEIL GLOVER, Rolls Royce PLC

PROFESSOR JOHN KILNER, Imperial College

Discussion

Session 2: Health and Environmental Effects of Novel Materials

Chair: PROFESSOR STEPHEN HOLGATE

PROFESSOR GUNTER OBERDÖRSTER, University of Rochester, USA

PROFESSOR VICKI COLVIN, Rice University, USA

PROFESSOR ANTHONY SEATON, University of Aberdeen

Discussion

Session 3: Governance and Regulation Issues

Chair: PROFESSOR SUSAN OWENS

PROFESSOR SHEILA JASANOFF, Harvard University, USA

PROFESSOR DANIEL SAREWITZ, Arizona State University, USA

Discussion

CORNELIS BREKELMANS, European Commission

DR RICHARD OWEN, Environment Agency

Discussion

In addition to the speakers and Members of the Commission, other participants included:

Name	Organisation
Professor George Attard	Southampton University
Mr David Bonser	British Nuclear Fuels PLC
Dr Rachel Brazil	Royal Society of Chemistry
Ms Anne Cassidy	British Standards Institution (BSI)
Mr Paul Collins	Bond Pearce
Mr Stuart Combes	Ministry of Defence
Dr Rob Doubleday	University of Cambridge
Dr Steve Fairhurst	Health and Safety Executive
Dr Stephen Feist	Centre for Environment, Fisheries and Aquaculture Science (CEFAS)
Dr Robin Fielder	Health Protection Agency
Dr Steffi Friedrichs	Nanotechnology Industries Association
Dr Elaine Groom	Queen's University Belfast
Dr Peter Hatto	IonBond Ltd
Dr Graham Holt	Cranfield University
Professor Kevin Jones	Lancaster University
Dr Peter Kearns	Organisation for Economic Co-operation and Development
Professor Frank Kelly	King's College, London
Dr Pamela Kempton	Natural Environment Research Council (NERC)
Dr Kerry Kirwan	Warwick University
Dr Tony Klepping	Begbroke Science Park, University of Oxford
Mr Daniel Lawrence	UK Environmental Law Association
Dr Qintao Liu	AstraZeneca
Professor Robert Maynard	Health Protection Agency
Mr Des McGraham	Thomas Swan and Co Ltd
Dr Mark Morrison	Institute of Nanotechnology
Dr Peter O'Neill	Department for International Development
Professor Nick Pidgeon	Cardiff University
Mr Richard Pitts	Office of Science and Innovation
Professor Jane Plant	Imperial College
Dr Rachel Quinn	The Royal Society

Name	Organisation
Professor Oliver Raymond	Centre of Excellence for Nano, Micro and Photonic Systems, Cenamps
Ms Marion Schulte zu Berge	Department for Environment, Food and Rural Affairs
Professor George Smith	University of Oxford
Professor Terry Tetley	Imperial College
Dr Eva Valsami-Jones	Natural History Museum
Dr John Wand	Engineering and Physical Sciences Research Council
Dr Jonathan Wentworth	Parliamentary Office of Science and Technology
Dr Angela Wilkinson	James Martin Institute for Science and Civilization, University of Oxford

Appendix D

MEMBERS OF THE ROYAL COMMISSION

CHAIRMAN

Professor Sir John Lawton CBE FRS

- President, Council of the British Ecological Society, 2005-2007

- Chief Executive, Natural Environment Research Council, 1999-2005

- Director (and founder), Natural Environment Research Council Centre for Population Biology at Imperial College, Silwood Park, 1989-1999

- Member, Royal Commission on Environmental Pollution, 1996-1999

- Lecturer, Senior Lecturer, Reader, Professor of Biology, University of York, 1972-1989

- Demonstrator in Animal Ecology, Department of Zoology, University of Oxford, 1968-1971

- Chairman, Royal Society for the Protection of Birds, 1993-1998

- Vice-President, Royal Society for the Protection of Birds, 1999-

- Past Vice-President, British Trust for Ornithology, 1999-2007

- Trustee, WWF-UK, 2002-2008; Fellow of WWF-UK, 2008-

- Foreign Associate, US National Academy of Sciences, 2008

- Foreign Honorary Member, American Academy of Arts and Sciences, 2008

MEMBERS

Professor Nicholas Cumpsty FREng

- Professor of Mechanical Engineering, Imperial College, 2005-2008

- Emeritus Professor of Mechanical Engineering, Imperial College, 2008-

- Member, Defence Science Advisory Council, 2005-

- Visiting Professor, Department of Aeronautics and Astronautics, Massachusetts Institute of Technology, 2005-

- Chief Technologist, Rolls-Royce plc, 2000-2005

- Lecturer, Reader, Professor, University of Cambridge, 1972-1999

- Director of the Whittle Laboratory, University of Cambridge, 1989-1999

Professor Michael H. Depledge DSc CBiol FIBiol FZS FRSA

- Professor of Environment and Human Health, Peninsula Medical School, Universities of Exeter and Plymouth

- Former Keeley Visiting Fellow, Wadham College, University of Oxford, 2006-2007

- Honorary Visiting Professor, Department of Zoology, University of Oxford

- Senior Science Advisor, Plymouth Marine Laboratory, 2005-2007

- Chief Scientific Advisor, Environment Agency of England and Wales, 2002-2006

- Vice-Chairman, Science Advisory Committee, European Commission, DG-Research, 2006-

- Board Member, Natural England, 2006-

- Council Member, Natural Environment Research Council, 2003-2006

- Honorary Professor, School of Earth Sciences and Engineering, Imperial College, 2002-

- Honorary Visiting Scientist, School of Public Health, Harvard University, USA, 2000-2003

Professor Paul Ekins

- Professor of Energy and Environment Policy, King's College, London, 2008-

- Head, Environment Group, Policy Studies Institute, 2000-2007

- Professor of Sustainable Development, University of Westminster, 2002-2007

- Chairman, National Industrial Symbiosis Programme

- Senior Consultant, Cambridge Econometrics

- Specialist Advisor, House of Commons Environmental Audit Committee, 1997-2005

- Member, Environmental Advisory Group, Ofgem

- Member, Sustainable Energy Policy Advisory Board, 2003-2007

- Trustee and Special Advisor, Right Livelihood Awards Foundation

- Chairman, Judging Panel, Ashden Awards for Sustainable Energy

- Trustee, Global Action Plan

Dr Ian Graham-Bryce CBE FRSC FRSE

- Principal Emeritus, University of Dundee

- Chairman, East Malling Trust for Horticultural Research

- Principal and Vice-Chancellor, University of Dundee, 1994-2000

- Convener, Committee of Scottish Higher Education Principals, 1998-2000

- President, Scottish Association for Marine Science, 2000-2004; and currently Honorary Vice-President

- President, British Crop Protection Council, 1996-2000

- Council Member, Natural Environment Research Council, 1989-1996

- Head, Environmental Affairs Division, Shell International, 1986-1994

- President, Association of Applied Biologists, 1988-1989

- Director, East Malling Research Station, 1979-1986

- President, Society of Chemical Industry, 1982-1984

Professor Stephen Holgate FRCP FMedSci FRSA

- Medical Research Council Clinical Professor of Immunopharmacology, University of Southampton

- Honorary Consultant Physician, Southampton University Hospital Trust

- Immediate Past President, British Thoracic Society

- Former Advisor, House of Lords Select Committee on Science and Technology

- Chairman, Expert Panel on Air Quality Standards (Department for Environment, Food and Rural Affairs (Defra))

- Chairman, Advisory Committee on Hazardous Substances (Defra)

- Seat on various Government advisory committees, including the Committee on the Medical Effects of Air Pollution (COMEAP) (Department of Health) and the Advisory Committee on Novel Foods and Processes (Food Standards Agency)

- Member, World Health Organization (WHO) Scientific Advisory Committee on Clean Air For Europe (CAFE), 2002-2004

- Chairman, Science in Health Group of the Science Council

- Chairman, Physiological Systems and Clinical Sciences Board, Medical Research Council

- Chairman of Medical Research Council Population Sciences and Medicines Board

- Member of Medical Research Council Strategy Board

Professor Jeffrey Jowell QC

- Professor of Law, University College London

- UK's Member on the Council of Europe's Commission for Democracy Through Law ("The Venice Commission")

- Chair, British Waterways Ombudsman Committee

- Non-executive Director of the Office of Rail Regulation, 2004-2007

- Practising barrister

Professor Peter Liss CBE FRS

- Professor of Environmental Sciences, University of East Anglia, 1985-

- Chair, Scientific Committee of the International Geosphere-Biosphere Programme (IGBP), 1993-1997

- Chair, International Scientific Steering Committee, Surface Ocean Lower Atmosphere Study (SOLAS), 2002-2007

- Council Member, Natural Environment Research Council, 1990-1995

- Independent Member, Inter-Agency Committee on Marine Science and Technology, 2000-2008

- Chair, Royal Society Global Environmental Research Committee, 2007-

- Council Member, Marine Biological Association of the UK

- Chair, Higher Education Funding Council's Research Assessment Exercise Panel in Earth and Environmental Sciences, 2001

- Guest Professor, Ocean University of Qingdao, China

- President, Challenger Society for Marine Science, 2006-2008

- Chair, European Research Council Advanced Grants Panel in Earth System Science, 2008-

Professor Susan Owens OBE AcSS FRSA FRGS HonMRTPI

- Professor of Environment and Policy, Department of Geography, University of Cambridge; and Professorial Fellow of Newnham College

- Honorary Professor, University of Copenhagen, 2008-

- King Karl XVI Gustaf Visiting Professor, Stockholm University and Royal Institute of Technology (KTH), 2008-2009

- Member, Sub-Panel H31 (Town and Country Planning), 2008 Research Assessment Exercise, 2005-

- Member, Strategic Research Board, Economic and Social Research Council, 2007-

- Member, Steering Committee, Office of Science and Innovation Review of Science in Department for Environment, Food and Rural Affairs, 2005-2006

- Member, Countryside Commission, 1996-1999

- Member, UK Round Table on Sustainable Development, 1995-1998

- Member, Deputy Prime Minister's Expert Panel during preparation of 1998 Transport White Paper, 1997-1998

Professor Judith Petts AcSS FRSA FRGS

- Pro-Vice-Chancellor (Research and Knowledge Transfer), University of Birmingham

- Chair of Environmental Risk Management, University of Birmingham

- Member, Engineering and Physical Sciences Research Council Societal Issues Panel

- Member, Environmental Advisory Board, Veolia Environmental

- Member, Higher Education Funding Council's Research Assessment Exercise Panel in Geography and Environmental Studies

- Member, DIUS Sciencewise Expert Resource Centre Steering Group

- Head, School of Geography, Earth and Environmental Sciences, University of Birmingham, 2002-2007

- Council Member, Natural Environment Research Council, 2000-2006

- Former Specialist Advisor, House of Commons Environment, Transport and Regional Affairs Committee and House of Lords Sub-Committee C

- Member, Council of the Institute of Environmental Assessment, 1990-2000

Professor Steve Rayner FRAI FRSA FAAAS FSfAA

- Director, James Martin Institute for Science and Civilization, Professor of Science and Civilization, Saïd Business School, University of Oxford; and Professorial Fellow of Keble College

- Professor of Environment and Public Affairs, Columbia University, USA, 1999-2003

- Chief Scientist, Pacific Northwest National Laboratory, USA, 1996-1999

- Director, Economic and Social Research Council Science in Society Programme, 2001-2007

- Member, Intergovernmental Panel on Climate Change

- Past President of the Sociology and Social Policy Section of the British Association

John Speirs CBE LVO

- Member of the Chemistry Leadership Council, 2004; and Chairman of its Futures Group Committee, 2003-2004

- Past President, National Society for Clean Air and Environmental Protection, 2002-2003

- Director, The Carbon Trust, 2001-2007

- Member, Management Committee of the Prince of Wales's Business and Environment Programme, 1997-2004; and Chairman of its UK Faculty, 1994-2002

- Member, Advisory Committee, Kleinwort Benson Equity Partners, 1991-2007

- Chairman, Dramgate Ltd, 1991-1995

- Managing Director, Norsk Hydro (UK) Ltd, 1981-2001

- Member of the Aluminium Federation Council, 1992-2002; and Past President, 1997-1998

- Council Member, Chemical Industries Association, 1993-2002; and Chairman of its Public Affairs Committee, 1993-2000

- Member, Science and Engineering Research Council, 1993-1994

- Member, Government's Advisory Committee on Business and the Environment, 1991-1995

- Chairman, Merton, Sutton and Wandsworth Family Health Services Authority, 1989-1995

- Divisional Director, The National Enterprise Board, 1976-1981

Professor Janet Sprent OBE FRSE FRSA FLS

- Emeritus Professor of Plant Biology, University of Dundee

- Honorary Research Fellow, Scottish Crop Research Institute

- Trustee, Royal Botanic Gardens Edinburgh

- Board Member, Scottish Natural Heritage, 2001-2007

- Member, Scottish Higher Education Funding Council, 1992-1996

- Council Member, Natural Environment Research Council, 1991-1995

- Governor, Macaulay Land Use Research Institute, 1990-2000; and Chairman, 1995-2000

- Honorary Member, British Ecological Society

- Fellow, Macaulay Land Use Research Institute

Professor Lynda Warren FiBIOL

- Emeritus Professor of Environmental Law, Aberystwyth University

- Deputy Chair, Joint Nature Conservation Committee

- Member, Committee on Radioactive Waste Management

- Board Member, British Geological Survey

- Board Member, Environment Agency, 2000-2006

- Chair, Wales Coastal and Maritime Partnership

- Chair, Salmon and Freshwater Fisheries Review, 1998-2000

- Former Member, Radioactive Waste Management Advisory Committee, 1994-2003

- Former Member, Countryside Council for Wales, 1991-2003

- Trustee, Field Studies Council

- Trustee, Wildlife Trust of South and West Wales

- Former Trustee, WWF-UK

Appendix E

Examples of properties of materials and nanomaterials

Description	Possible environmental significance
Crystal structure A material is crystalline if the atoms from which it is made up are organised into a repeating pattern over long distances. The properties of a crystalline material can be different to those of a non-crystalline analogue, and some materials can have more than one crystalline state.	The crystallinity of a substance can affect its properties, which could lead to different biological effects. For example, crystalline silica is more harmful if inhaled than amorphous silica.
Magnetism Some materials display strong magnetic properties, which can be desirable (e.g. for information storage).	There is some evidence that magnetic nanoparticles exist naturally in some organisms, purportedly for navigational purposes.
Electrical conductivity Some materials conduct electricity, others are insulators; semiconductors are of particular interest for the electronics industry.	It has not been suggested to us that this property presents a problem either for the environment or human health.

Description	Possible environmental significance
Defects/impurities/dopants The properties of some materials can be substantially altered by defects or imperfections in their structure, indeed the effects of defects can be deliberately sought. A defect could be an absence of an atom from the crystal structure or the inclusion of an additional atom in the structure, both of which could alter the mechanical, electrical or chemical properties of the material. Crystal defects can also include point, line and planar defects (e.g. vacancies, dislocations, grain boundaries, etc.), each of which plays a different role. Impurities can result from poor quality control in the preparation of a material, and will have some impact on its properties. Alloys, such as those of aluminium, are substances in which an impurity has been added to a pure metal to modify deliberately its properties. In other materials, an atom can be included at specific points of the crystal structure to modify properties; this could be considered as a dopant (a deliberate impurity) which is common in semiconductors.	The change in properties could affect the behaviour of the materials in the environment, for example, manganese doped titanium dioxide is used in sunscreen, and generates fewer free radicals than its undoped equivalent.
Porosity Some materials (such as zeolites) contain cavities or pores as part of their structure, dramatically increasing the available surface area, which can change the reactivity of the material or its ability to bind to another substance. Some porous materials are proposed as a means of binding and storing hydrogen; others are used to separate one substance from another.	Porous materials could provide a means of transporting molecules through the environment, or could adsorb materials from the environment. This may have an effect in itself, or could lead to an effect if a porous material is ingested.
Sorption The binding of a gas or liquid to the surface of a material is adsorption, whereas the diffusion of a substance into the bulk volume of a material is absorption. The sorption of one substance to another can alter the properties of the absorber/adsorber and can provide a mechanism for the transport of that which is sorbed. Sorption can also have a major effect on catalytic activity, notably by poisoning active sites.	If a material absorbs or adsorbs another substance, this could affect its mobility in the environment and the bioavailability of the sorbed substance. It could also have an effect if the material is ingested.

Description	Possible environmental significance
Surface chemistry	
For reactions to take place at the surface of a material, active sites, i.e. places where the chemistry can occur, must be present.	Substantial surface activity may give rise to pronounced bioactivity.
Surface charge	
This is the electric charge that is found on the surface of a material, be it a particle, a protein in solution or a solid semiconductor. It is an important parameter in determining the chemistry and interactions of a material, particularly how it forms solutions or suspensions.	How a material interacts with the environment will determine its mobility in the environment and its bioavailability.
Hydrophobicity/hydrophilicity	
The repellent or attractive behaviour of a material towards water is related to the surface energy of the material, and can be quantified by measuring the contact angle of a water droplet with a flat surface of the material.	How hydrophilic/hydrophobic a material is will have a significant effect on its mobility in the environment and its bioavailability.
State of aggregation	
Small particles can either be evenly dispersed throughout a medium, or they can clump together and aggregate (i.e. strongly adhere together). Hydrophobic particles in water will aggregate rapidly, whereas hydrophilic particles in water will disperse evenly. In order to control the behaviour of particles, surface modifications can be made to overcome undesired behaviour, for example to prevent hydrophobic particles aggregating in water one could attach surfactants to the particle surface. The aim of such surface modification is to alter the properties of a material so that they can be displayed in an environment where these materials would not otherwise be found. In addition to modifying the environment in which a material may be found, these surface modifications could have their own properties which could alter the mobility of a material and its toxicity.	The state of aggregation has an effect on the properties of a material, in particular on how it is transported through the environment, its bioavailability and toxicity.

Appendix F

SOLUTIONS AND DISPERSIONS

Solution: In which one substance (the solute) is homogeneously distributed through (dissolved in) a solvent. The solute can be ionic, molecular or atomic.

Dispersion: The distribution of one substance through another without being dissolved.

Suspension: A dispersion of particles (usually greater than 1,000 nm in size) through a fluid, for example particles of mud suspended in river water. Many suspensions will settle if left undisturbed.

Colloid: A mixture of one substance dispersed evenly throughout another, consisting of a dispersed phase and continuous phase. The dispersed and continuous phases can be solid, liquid or gas. The size of the dispersed phase particles is usually 1-1,000 nm. Colloids can be unstable and can suffer from sedimentation or aggregation of the dispersed medium leading to separation of the phases. Stabilisation of colloids can be achieved by maintaining the separation between particles of the dispersed phase, thus preventing aggregation. The normal stabilisation methods are through the addition of a stabilising agent (such as a polymer) that helps maintain the even dispersion, or by increasing the electrostatic repulsion of the particles of the dispersed phase. It should be noted that colloids can also be destabilised by the addition of substances that disrupt the separation of the dispersed phase or the properties of the continuous phase.

Appendix G

Dust-related lung disease

G.1 Most diffuse fibrotic lung disease is of unknown origin, but in an occupational setting, entrapment of inhaled mineral dusts in the periphery of the lungs (alveolar air sacs and small airways) is an important cause. These dusts include naturally-occurring nanoparticles and nanoparticles produced inadvertently in other processes such as drilling. Particle toxicity is a function of size, composition, dose, duration of exposure and susceptibility, causing a combination of chronic inflammation and aberrant repair. In coal miners this leads to pneumoconiosis, and in stone workers and sand blasters to silicosis, causing the lungs to become stiff leading to breathlessness and impaired gas exchange.

G.2 Heavy exposure to asbestos fibres used as insulation and as a fire retardant is also a cause of lung fibrosis (asbestosis) with amosite and crocidolite being more fibrogenic than either chrysotile or tremolite.[1] Asbestos exposure can also lead to pleural plaques that are visible by X-ray when they calcify and are not linked to mesothelioma, although extensive diffuse pleural fibrosis occasionally occurs.[2]

G.3 Asbestos exposure in miners, shipbuilders and plumbers is also the cause of a highly malignant cancer of the outer lining of the lungs (mesothelioma) and increases the risk of lung cancer. Here, malignancy occurs at very low fibre concentrations and can occur in family members of asbestos workers exposed to asbestos on clothing. Mesothelioma is most strongly associated with a combination of both fibre width and length.[3,4] Crocidolite is the most active form of asbestos fibre, possibly acting by inducing DNA mutations through nitric oxide induction.[5] Size, chemical form and shape are the critical features determining risk of mesothelioma – either short, high or chronic, low exposure patterns being sufficient to produce tumours many years later.[6]

G.4 An important and treatable form of fibrotic lung disease is extrinsic allergic alveolitis (EAA, hypersensitivity pneumonitis) caused by inhalation of biological particles such as proteins in the droppings of birds especially pigeons and budgerigars (bird fancier's lung), actinomycetes from mouldy hay (farmer's lung) and mould in compost (mushroom picker's lung).[7] The organic particles stimulate a complex and vigorous immune response in the lung leading to inflammation and fibrosis. In its active inflammatory phase, EAA is responsive to the inhibitory actions of corticosteroids; this is not the case for mineral dust-induced fibrosis.

G.5 Improvements in occupational protection have now made both of these forms of lung fibrosis uncommon, although the incidence of mesothelioma will continue to increase for about 15 years, in relation to known periods of prior asbestos exposure.[8]

REFERENCES

1   Churg, A., Wright, J., Wiggs, B. and Depaoli, L. (1990). Mineralogic parameters related to amosite asbestos-induced fibrosis in humans. *Am. Rev. Respir. Dis.*, **142**(6 Part 1), 1331-1336.

2   Gevenois, P.A., de Maertelaer, V., Madani, A., Winant, C., Sergent, G. and De Vuyst, P. (1998). Asbestosis, pleural plaques and diffuse pleural thickening: Three distinct benign responses to asbestos exposure. *Eur. Respir. J.*, **11**(5), 1021-1027.

3   Churg, A. and Wiggs, B. (1984). Fiber size and number in amphibole asbestos-induced mesothelioma. *Am. J. Pathol.*, **115**(3), 437-442.

4   Churg, A. and Wiggs, B. (1987). Accumulation of long asbestos fibers in the peripheral upper lobe in cases of malignant mesothelioma. *Am. J. Ind. Med.*, **11**(5), 563-569.

5   Roggli, V.L. (1995). Malignant mesothelioma and duration of asbestos exposure: Correlation with tissue mineral fibre content. *Ann. Occup. Hyg.*, **39**(3), 363-374.

6   Unfried, K., Schürkes, C. and Abel, J. (2002). Distinct spectrum of mutations induced by crocidolite asbestos: Clue for 8-hydroxydeoxyguanosine-dependent mutagenesis *in vivo*. *Cancer Res.*, **62**(1), 99-104.

7   Woda, B.A. (2008). Hypersensitivity pneumonitis: An immunopathology review. *Arch. Pathol. Lab. Med.*, **132**(2), 204-205.

8   British Thoracic Society (BTS) Standards of Care Committee (2007). BTS statement on malignant mesothelioma in the UK, 2007. *Thorax*, **62**(Suppl. 2), ii1-ii19.

Appendix H

ADVERSE HEALTH EFFECTS OF PARTICULATE AIR POLLUTION

H.1 From extensive epidemiological studies worldwide there has been increasing concern over the contribution of inhaled vehicle-derived particulates from exhaust emissions to adverse health outcomes, especially asthma, chronic obstructive pulmonary disease, pneumonia and cardiovascular disease including stroke.[1]

H.2 Primary particles are generated during combustion and comprise a mixture of elemental carbon, volatile organic chemicals and metals. Secondary particles are produced in the atmosphere from gaseous pollutants such as oxides of nitrogen (NO_x) and sulphur dioxide (SO_2) to form nitrate and sulphate salts respectively.[2]

H.3 While much of the particulate mass is in the size fraction 10 μm mass median aerodynamic diameter (MMD) or above (PM_{10}), by far the greatest numbers of particles fall into the fine range of 0.25 μm ($PM_{0.25}$) or less than 0.1 μm ($PM_{0.1}$), i.e. nanoparticles, with the number increasing exponentially with falling MMD.[3] Such particles preferentially deposit in the alveoli at the lung periphery or are exhaled.[4]

H.4 Deposited pollutant particles are taken up by luminal macrophages, but this disposal mechanism becomes easily saturated by submicronic particles,[5] leaving free particles to be absorbed through the alveolar epithelium by transcellular or paracellular mechanisms.

H.5 Depending on the chemical and physical nature of pollutant particles, their uptake by macrophages and epithelial cells stimulates the release of inflammatory mediators[6] and interferes with innate immunity to reduce defence against infection.[7]

H.6 In the UK particulate air pollution accounts for up to 24,000 premature deaths from cardiovascular disease a year including heart attacks and stroke.[8] Although not known for certain, possible mechanisms include increased cardiac arrhythmia, activation of the clotting pathways, increased platelet stickiness and enhanced inflammation of atheromatous plaques in arteries.[9]

H.7 Ultrafine particles ($PM_{1-2.5}$) account for the majority of the cardiovascular events,[10] fresh primary particles and those from diesel emissions being most active.[11]

REFERENCES

1   Chen, Y., Craig, L. and Krewski, D. (2008). Air quality risk assessment and management. *J. Toxicol. Environ. Health A*, **71**(1), 24-39.

2   Reiss, R., Anderson, E.L., Cross, C.E., Hidy, G., Hoel, D., McClellan, R. and Moolgavkar, S. (2007). Evidence of health impacts of sulfate- and nitrate-containing particles in ambient air. *Inhal. Toxicol.*, **19**(5), 419-449.

3   de Hartog, J.J., Hoek, G., Mirme, A., Tuch, T., Kos, G.P., ten Brink, H.M., Brunekreef, B., Cyrys, J., Heinrich, J., Pitz, M., Lanki, T., Vallius, M., Pekkanen, J. and Kreyling, W.G. (2005). Relationship between different size classes of particulate matter and meteorology in three European cities. *J. Environ. Monit.*, **7**(4), 302-310.

4   Asgharian, B. and Price, O.T. (2007). Deposition of ultrafine (nano) particles in the human lung. *Inhal. Toxicol.*, **19**(13), 1045-1054.

5   Muhlfeld, C., Rothen-Rutishauser, B., Blank, F., Vanhecke, D., Ochs, M. and Gehr, P. (2008). Interactions of nanoparticles with pulmonary structures and cellular responses. *Am. J. Physiol. Lung Cell. Mol. Physiol.*, **294**, L817-L829.

6   Ishii, H., Hayashi, S., Hogg, J.C., Fujii, T., Goto, Y., Sakamoto, N., Mukae, H., Vincent, R. and van Eeden, S.F. (2005). Alveolar macrophage-epithelial cell interaction following exposure to atmospheric particles induces the release of mediators involved in monocyte mobilization and recruitment. *Respir. Res.*, **6**, 87.

7   Lundborg, M., Dahlén, S.E., Johard, U., Gerde, P., Jarstrand, C., Camner, P. and Låstbom, L. (2006). Aggregates of ultrafine particles impair phagocytosis of microorganisms by human alveolar macrophages. *Environ. Res.*, **100**(2), 197-204.

8   Committee on the Medical Effects of Air Pollutants (COMEAP) (1998). *The quantification of the effects of air pollution on health in the United Kingdom.* HMSO, London.

9   Mills, N.L., Törnqvist, H., Robinson, S.D., Gonzalez, M.C., Söderberg, S., Sandström, T., Blomberg, A., Newby, D.E. and Donaldson, K. (2007). Air pollution and atherothrombosis. *Inhal. Toxicol.*, **19**(Suppl. 1), 81-89.

10 Duffin, R., Mills, N.L. and Donaldson, K. (2007). Nanoparticles – A thoracic toxicology perspective. *Yonsei Med. J.*, **48**(4), 561-572.

11 Mills, N.L., Törnqvist, H., Gonzalez, M.C., Vink, E., Robinson, S.D., Söderberg, S., Boon, N.A., Donaldson, K., Sandström, T., Blomberg, A. and Newby, D.E. (2007). Ischemic and thrombotic effects of dilute diesel-exhaust inhalation in men with coronary heart disease. *N. Engl. J. Med.*, **357**(11), 1075-1082.

Appendix I

MECHANISM OF ENTRY OF NANOPARTICLES INTO EPITHELIAL CELLS

I.1   Nanoparticles are able to pass into lung epithelial cells both through specific energy-dependent and non-dependent mechanisms, which are highly dependent on particle size, shape and surface characteristics.[1]

I.2   When particles are coated with proteins, peptides or charged polymers to change their surface properties, this enables them to enter the cell's cytoplasm and nucleus.[2]

I.3   Manufactured nanoparticles are treated by cells as invading micro-organisms utilising similar cell uptake and intracellular transport machinery.

I.4   Recent studies show that a fraction of the nanoparticles that pass into lung tissue are subsequently transported to the larynx via the lymphatics to be re-secreted onto the surface epithelium for removal.[3]

I.5   There remains considerable uncertainty over the ability of manufactured nanoparticles to cross epithelial surfaces and become systemically available. This uncertainty relates to wide inter-species differences in the structure and function of epithelial barriers,[4] differences between the handling of instilled versus inhaled particles in the lung and differences in the ability of particles to pass into epithelial cells.

REFERENCES

1   Leonenko, Z., Finot, E. and Amrein, M. (2007). Adhesive interaction measured between AFM probe and lung epithelial type II cells. *Ultramicroscopy*, **107**(10-11), 948-953.

2   Fuller, J.E., Zugates, G.T., Ferreira, L.S., Ow, H.S., Nguyen, N.N., Wiesner, U.B. and Langer, R.S. (2008). Intracellular delivery of core-shell fluorescent silica nanoparticles. *Biomaterials*, **29**(10), 1526-1532.

3   Semmler-Behnke, M., Takenaka, S., Fertsch, S., Wenk, A., Seitz, J., Mayer, P., Oberdörster, G. and Kreyling, W.G. (2007). Efficient elimination of inhaled nanoparticles from the alveolar region: Evidence for interstitial uptake and subsequent reentrainment onto airways epithelium. *Environ. Health Perspect.*, **115**(5), 728-733.

4   Bermudez, E., Mangum, J.B., Wong, B.A., Asgharian, B., Hext, P.M., Warheit, D.B. and Everitt, J.I. (2004). Pulmonary responses of mice, rats, and hamsters to subchronic inhalation of ultrafine titanium dioxide particles. *Toxicol. Sci.*, **77**(2), 347-357.

Appendix J

CURRENT REGULATIONS THAT AFFECT NANOMATERIALS[1]

Legislation	Consumer Protection	Health & Safety	Environmental Protection
Notification of New Substances Regulations 1993		X	
Registration, Evaluation, Authorisation and Restriction of Chemicals (REACH)		X	
Biocidal Products Regulations 2001 (as amended)	X		
Chemicals (Hazard Information and Packaging for Supply) Regulations 2002 (as amended)	X	X	
Control of Major Accident Hazard Regulations 1999 (as amended)		X	X
Control of Substances Hazardous to Health Regulations 2002 (as amended)		X	
Dangerous Substances & Explosions Atmosphere Regulations 2002		X	
Health & Safety at Work Act 1974		X	
Management of Health & Safety at Work Regulations		X	
Ammonium Nitrate Materials (High Nitrogen Content) Safety Regulations 2003			X
Batteries and Accumulators (Containing Dangerous Substances) Regulations 1994 (as amended)	X		X
Medical Devices Regulations 2002 (as amended)	X		X
Medicines Act 1968	X		X
Medicines for Human Use (Marketing Authorisations etc.) Regulations 1994 (as amended)	X		
Motor Fuel (Composition and Content) Regulations 1999 (as amended)	X		
End-of-Life Vehicles Regulations 2003			X
Restriction of the Use of Certain Hazardous Substances in Electrical and Electronic Equipment Regulations 2005			X

Legislation	Consumer Protection	Health & Safety	Environmental Protection
Directive 2002/96/EC on Waste Electrical and Electronic Equipment			X
Packaging (Essential Requirements) Regulations 2003			X
Producer Responsibility Obligations (Packaging Waste) Regulations 2005			X
Veterinary Medicines Regulations 2005	X		
Building Regulations 2000 (as amended)	X		
Textile Products (Indications of Fibre Content) Regulations 1986 (as amended)	X		
Electrical Equipment (Safety) Regulations 1994	X		
Control of Pesticides Regulations 1986 (as amended)	X	X	
Fertilisers Regulations 1991 (as amended)	X		
Plant Protection Products Regulations 2005 (as amended)	X		
Detergents Regulations 2005	X		
Cosmetic Products (Safety) Regulations 2004 (as amended)	X		
General Product Safety Regulations 2005	X		
Additives Directive 89/107/EEC (as amended)	X		
Articles in Contact with Food Regulations 1987 (as amended)	X		
Colours in Food Regulations 1995 (as amended)	X		
Contaminants in Food (England) Regulations 2005	X		
Food Safety Act 1990 (as amended)	X		
Materials and Articles in Contact with Food (England) Regulations 2005	X		
Miscellaneous Food Additives Regulations 1995 (as amended)	X		
Novel Foods and Novel Food Ingredients Regulations 1997 (as amended)	X		
Plastic Materials and Articles in Contact with Food Regulations 1998 (as amended)	X		
Regulation (EC) No. 178/2002 on General Principles of Food Law	X		

Legislation	Consumer Protection	Health & Safety	Environmental Protection
Environmental Protection Act 1990 (as amended)			X
Pollution Prevention and Control (England and Wales) Regulations 2000 (as amended)			X
Control of Pollution (Oil Storage) (England) Regulations 2001			X
Control of Pollution (Silage, Slurry and Agricultural Fuel Oil) Regulations 1991 (as amended)			X
Environmental Protection (Prescribed Processes and Substances) Regulations 1991			X
Air Quality (England) Regulations 2000			X
Clean Air Act 1993			X
Air Quality Limit Values Regulations 2003			X
Groundwater Regulations 1998			X
Surface Waters (Dangerous Substances) (Classification) Regulations 1997			X
Surface Waters (Dangerous Substances) (Classification) Regulations 1998			X
Trade Effluents (Prescribed Processes and Substances) Regulations 1989 (as amended)			X
Urban Waste Water Treatment (England and Wales) Regulations 1994			X
Waste Management Licensing Regulations 1994			X
Water Act 2003			X
Water Environment (Water Framework Directive) (England and Wales) Regulations 2003			X
Water Industry Act 1991		X	X
Water Resources Act 1991		X	X
Hazardous Waste (England and Wales) Regulations 2005			X
Landfill (England and Wales) Regulations 2002			X
List of Wastes (England) Regulations 2005			X
Waste Incineration (England and Wales) Regulations 2002			X

131

REFERENCES

1   Frater, L., Stokes, E., Lee, R. and Oriola, T. (2006). *An overview of the framework of current regulation affecting the development and marketing of nanomaterials.* A report for the Department for Trade and Industry (DTI). December 2006.

ABBREVIATIONS

BERR	Department for Business, Enterprise and Regulatory Reform
BRASS	Economic and Social Research Council (ESRC) Centre for Business Relationships, Accountability, Sustainability and Society at Cardiff University
C_{60}	carbon-60
CHIP	Chemicals (Hazard Information and Packaging for Supply) Regulations 2002
CSL	Central Science Laboratory
CTA	Constructive Technology Assessment
Defra	Department for Environment, Food and Rural Affairs
DG SANCO	Directorate General for Health and Consumer Affairs
DH	Department of Health
DIUS	Department for Innovation, Universities and Skills
DOENI	Department of Environment, Northern Ireland
EA	Environment Agency
ECA	European Chemicals Agency
EEA	European Environment Agency
EHS	environmental, health and safety data
EINECS	European Inventory of Existing Chemical Substances
ENP	engineered nanoparticle
EPA	Environmental Protection Agency, United States
EPSRC	Engineering and Physical Sciences Research Council
EU	European Union
FDA	Food and Drug Administration, United States
FlFFF-ICP-MS	flow field-flow fractionation inductively coupled plasma mass spectrometry
FSA	Food Standards Agency
HSE	Health and Safety Executive
I^{125}	iodine-125
IPPC	Integrated Pollution Prevention and Control
IT	information technology
LCA	life cycle assessment
LCD	liquid crystal display

LDL	low density lipoprotein
μm	micrometre
METI	Ministry of Economy, Trade and Industry, Japan
MEXT	Ministry of Education, Culture, Sports, Science and Technology, Japan
MN	manufactured nanomaterials
MOD	Ministry of Defence
NATO	North Atlantic Treaty Organization
NEG	Nanotechnology Engagement Group
NERC	Natural Environment Research Council
NGOs	non-governmental organisations
NI	Northern Ireland
NIDG	Nanotechnology Issues Dialogue Group
nm	nanometre
NP	nanoparticle
NRCG	Nanotechnology Research Co-ordination Group
OECD	Organisation for Economic Co-operation and Development
OECD WPNM	Organisation for Economic Co-operation and Development Working Party on Manufactured Nanomaterials
PNEC	predicted no effects concentration
POPs	persistent organic pollutants
QD	quantum dot
QSAR	quantitative structure–activity relationship
R&D	research and development
RAE	Royal Academy of Engineering

REACH	Regulation (EC) No. 1907/2006 of the European Parliament and of the Council of 18 December 2006 concerning the Registration, Evaluation, Authorisation and Restriction of Chemicals (REACH), establishing a European Chemicals Agency, amending Directive 1999/45/EC and repealing Council Regulation (EEC) No. 793/93 and Commission Regulation (EC) No. 1488/94 as well as Council Directive 76/769/93 and Commission Directives 91/155/EEC, 93/67/EEC, 93/105/EC and 2000/21/EC
RS	Royal Society
RTTA	Real-Time Technology Assessment
SCENIHR	Scientific Committee on Emerging and Newly Identified Health Risks
SEPA	Scottish Environment Protection Agency
SPRU	Science and Technology Policy Research Unit at Sussex University
SWCNTs	single-walled carbon nanotubes
Tc^{99m}	technetium-99
TGD	Technical Guidance Document
TSCA	Toxic Substances Control Act, United States
UN	United Nations
US	United States of America
UK	United Kingdom
UV	ultra-violet
WEEE	Directive 2002/96/EC of the European Parliament and of the Council of 27 January 2003 on Waste Electrical and Electronic Equipment

Index

136

139

143

Printed in the UK by The Stationery Office Limited
on behalf of the Controller of Her Majesty's Stationery Office
ID 5799999   11/08

Printed on Paper containing 75% recycled fibre content minimum.